INFORMATION TECHNOLOGIES IN THE MINERALS INDUSTRY

PROCEEDINGS OF THE FIRST INTERNATIONAL CONFERENCE ON
INFORMATION TECHNOLOGIES IN THE MINERALS INDUSTRY VIA
THE INTERNET/1-12 DECEMBER 1997

Information Technologies in the Minerals Industry

Edited by

G. N. PANAGIOTOU & T. N. MICHALAKOPOULOS
National Technical University of Athens, Greece

A.A.BALKEMA/ROTTERDAM/BROOKFIELD/1998

CR-Rom with the complete text of the *Proceedings of the First International Conference on Information Technologies in the Minerals Industry* is enclosed in the back cover

Published by
A.A. Balkema, P.O. Box 1675, 3000 BR Rotterdam, Netherlands
Fax: +31.10.4135947; E-mail: balkema@balkema.nl; Internet site: http://www.balkema.nl

A.A. Balkema Publishers, Old Post Road, Brookfield, VT 05036-9704, USA
Fax: 802.276.3837; E-mail: info@ashgate.com

ISBN 90 5410 932 7

© 1998 A.A. Balkema, Rotterdam
Printed in the Netherlands

CONTENTS

V

Technical session II: Mine planning – Mining operations

Technical session III: Rock mechanics – Excavation engineering

Technical session IV: Mine equipment

Technical session V: Mine safety – Training

Technical session VI: Reclamation – Environmental issues

FOREWORD

The First International Conference on Information Technologies in the Minerals Industry - MineIT '97 was held, via the Internet, from 1-12 December 1997. It was organized by the Department of Mining Engineering and Metallurgy of the National Technical University of Athens (NTUA), Greece, but the conference's venue was obviously the Cyberspace. MineIT '97 was run using a web-pages arrangement set-up and maintained by the Mining Systems Simulation Unit (MSSLab) at NTUA.

During this period, the conference's web-site provided a forum for the presentation, discussion and criticism of state-of-the-art and emerging information technologies applied in the minerals industry, covering a wide spectrum of applications from orebody modelling to training and reclamation.

MineIT '97 was a 'virtual conference', but it had all the elements of a conventional conference, with a registration 'desk' (a personalised conference badge was sent to each registrant via the Internet upon registration), welcoming addresses, technical sessions etc., even a tour-sightseeing to the Acropolis of Athens. Over the conference period, participants were able to use the conference's web-site to tune in, or tune out and in again, many times, any time; to read and download papers, to 'write in' their comments and queries. Papers' authors defended their theses and offered comments to queries or criticisms. Following a number of requests by participants and members of the International Organizing Committee, the conference period was extended till the end of the year. During the course of the conference there were 1683 'hits' and 296 registrants, from 35 countries, submitted a non-compulsory registration form.

A wide range of high quality papers from North and South America, Europe, Australia, Africa and Asia were attracted. Forty-six papers were accepted for presentation and arranged in six technical sessions: Exploration – Orebody Modelling; Mine Planning – Mining Operations; Rock Mechanics – Excavation Engineering; Mine Equipment; Mine Safety – Training; Reclamation – Environmental Issues. Several interesting case histories were offered, while high quality 3D colour graphics and animations supported many papers.

This volume of proceedings is the permanent record of the 'virtual conference' and contains three keynote papers and the titles, abstracts and keywords of all the papers presented in MineIT '97. The companion CD-ROM includes not only the complete papers (most of them cannot be printed due to the complex colour-graphics they contain), but the whole conference's web-site pages.

The organization and success of a conference of this sort is due mainly to the timeless efforts and expertise of many individuals, authors included. Special thanks go to the sponsor of this conference O&K Orenstein and Koppel Inc., USA for its generous support and to the members of the International Organizing Committee. Particular recognition is accorded to Ms. Margarita Patesti, Mr. Kostas Kladis and Mr. Leonidas Lymberopoulos for their dedication and tireless efforts to prepare and maintain the conference's web-site.

George N. Panagiotou
Theodore N. Michalakopoulos
Editors

INTERNATIONAL ORGANISING COMMITTEE

Chair

Dr. George N. Panagiotou
National Technical University of Athens, Greece
panagiotou@metal.ntua.gr

Co-Chairs

Dr. Kostas Fytas
Universite Laval, Quebec, Canada
Kostas.Fytas@gmn.ulaval.ca

Dr. Uday Kumar
Lulea University of Technology, Sweden
uday.kumar@ce.luth.se

Members

Prof. Sukumar Bandopadhyay
University of Alaska Fairbanks, USA
ffs0b@aurora.alaska.edu

Mr. Detlef F. Bartsch
O&K Orenstein & Koppel Inc, USA
defraba@atl.mindspring.comc

Prof. Roussos Dimitrakopoulos
Bryan Research Centre, University of
Queensland, Australia
roussos@minmet.uq.oz.au

Paul Griffin
Surpac Software International Ltd.,
U.K.
paul@ssiuk.demon.co.uk

Stephen Henley
RCI Ltd / Snowden Associates
(Europe), UK
steve@rci.co.uk

Prof. Zbigniew J. Hladysz
South Dakota School of Mines and
Technology, USA
zhladysz@silver.sdsmt.edu

Prof. Zhenqi Hu
China University of Mining and
Technology, China
huzq@dns.cumt.edu.cn

Dr. Michael Kaas
Department of the Interior, USA
michael_kaas@ios.doi.gov

Dr. Yuri Kaputin
Oxus Mining Ltd., Russia
Yuri@oxuspenj.com.uz

Dr. Antoni Kidybinski
Central Mining Institute, Poland
btxak@boruta.gig.katowice.pl

Dr. Peter Knights
Pontificia Universidad Catolica de
Chile-Santiago, Chile
knights@ing.puc.cl

Prof. Basil Maglaris
National Technical University of
Athens, Greece
maglaris@ntua.gr

Prof. Per Nicolai Martens
RWTH Aachen, Germany
martens@bbk1.rwth-aachen.de

Dr. John A. Meech
University of British Columbia,
Canada
jam@mining.ubc.ca

Prof. Ramirez Oyanguren
Universidad Politechnica de Madrid,
Spain
pramirez@dexmi.upm.es

Dr. Jacek Paraszczak
Universite Laval, Quebec, Canada
Jacek.Paraszczak@gmn.ulaval.ca

Prof. A. Gunhan Pasamechmetoglu
Middle East Technical University,
Turkey
gunhan@rorqual.cc.metu.edu.tr
Prof. Vladimir Pavlovic
University of Belgrade, Yugoslavia
paja@rgf.org.yu

Dr. Vladimir Petros
VSB Technical University Ostrava,
Czech Republic
vladimir.petros@vsb.cz

Prof. Lev Puchkov
Moscow State Mining University,
Russia
rector@techno.dialnet.ru
ekuzmin@aha.ru

Dr. Andy M. Robertson
Robertson GeoConsutants Inc., Canada
andyr@info-mine.com

Dr. Damian Schofield
University of Nottingham, UK
enzds@unix.ccc.nottingham.ac.uk

Dr. Raj Singhal
Federal Government of Canada,
Canada
Singhal@agt.net

Gordon Smith
CSIR-Mining Technology, South
Africa
glsmith@csir.co.za

Dr. John Sturgul
University of Idaho, USA
sturgul@uidaho.edu

Dr. Ertugrul Topuz
Virginia Tech, USA
topuz@mail.vt.edu

Dr. Nick Vagenas
Laurentian University, Ontario, Canada
nvagenas@nickel.laurentian.ca

Prof. F. Ludwig Wilke
Technical University of Berlin,
Germany
wilke@bg.tu-berlin.de

Prof. Tuncel M. Yegulalp
Columbia University, USA
yegulalp@columbia.edu

Dr. Jon Yingling
University of Kentucky, USA
jyinglin@engr.uky.edu

The companion CD-ROM contains the web-pages of MineIT '97, including all papers, as presented during the conference's period. Certain modifications have been made in order to facilitate the post-symposium viewing of the web-pages.

To use the CD-ROM the minimum hardware requirements are: a PC running Windows 95 and a CD-ROM drive.

To view the conference's pages you need a web browser – for best viewing use the latest version of either Netscape Navigator or Microsoft Internet Explorer. If a web browser is not installed on your PC, the following software products have been downloaded, for your convenience, from the official web sites of the developers and are included in the CD-ROM under the folder 'browsers':

1. Netscape Navigator 4.

   To install run

   ```
   n32e404.exe
   ```

   in subfolder "netscape"

2. Microsoft Internet Explorer 4.

   To install run

   ```
   i34setup.exe
   ```

   in subfolder 'microsoft'

It is not necessary to have a connection to the Internet or to open a session with your Internet Service Provider (ISP) in order to view MineIT '97. You use the web browser simply for viewing the conference's pages contained in the CD-ROM.

Certain conference's pages have direct links to the Internet, e.g. the addresses of the e-mail boxes of papers' authors, organizations' web sites, etc. If you are viewing the pages while your PC is connected to the Internet, then these links are active, otherwise they will remain silent.

Insert the CD-ROM in the drive and open your web browser. Type in the URL address box:

```
[your CD-ROM drive letter, e.g. D or E]:\index.html
```

Now you are ready to move around the MineIT '97 pages. Don't forget to read the 'Read Me First' section first!

Note: The 'Send Message' option of the Navigation Bar on the top of each paper has been deactivated.

Keynote Papers

OPEN SYSTEMS STANDARDS FOR THE MINING INDUSTRY

P.F. Knights
Assistant Professor, Mining Centre, Catholic University of Chile

L.K. Daneshmend
Noranda/Falconbridge Chair in Mine-Mechanical Engineering, Queen's University, Canada

ABSTRACT: Key issues in managing information systems in the minerals industry will likely continue to be those relating to the integration and interoperability of disparate systems. This paper argues that, in order to harness the promised benefits of information technologies, the mining industry should adopt a set of open systems standards to define data formats and protocols for seamless data exchange. This is in direct contrast to the present situation where mining equipment, software and instrument suppliers rigidly adhere to proprietary standards for fear of losing competitive advantage. Benefits to the mining sector resulting from the adoption of such standards would be: the provision of near-time data for executive decision support, and the freedom to choose best technologies. The principal benefits to the leading mining software suppliers will be increased market share as a result of 'captive sites' converting to open systems standards. Examples are given of open systems standards developed for related industries such as the Petroleum Open Systems Corporation (POSC) and the Machinery Information Management Opens Systems Alliance (MIMOSA).

KEYWORDS: mining, open systems, mine management, data warehouse

1 INTRODUCTION

Software used by the mining industry reflects a mixture of generic industry applications and applications specific to mining. The generic application consist of spreadsheets, databases, word processing, CAD and GIS systems, maintenance management systems (including inventory and purchasing management), accounting, payroll and human resource management systems, project management systems and process control applications. The specialized mining applications include geotechnical and geological modelling, mine planning, mine dispatch and production control systems. It was recently estimated that 80% of the market for software in the mining industry is for purchase and support of generic applications. 20 % is for purchase and support of specialized applications. The conclusion for this reliance by mining companies on non-mining specific software is that the specialist software providers have failed to address issues such as product standards, flexibility and compatibility (Rychkun & Higgins, 1996).

3

2 STATEMENT OF THE PROBLEM

Information is usually synthesized from various data sources. In addition, the value of information decays with respect to time. In the face of the dynamic nature of today's commodity markets, mining companies are increasingly demanding ready access to historical databases in order to make informed decisions. Data which describes the status of a mining operation is collected and stored in diverse databases, not all of which use the same data structures, query languages and reporting formats. For example, if a senior executive notices that there has been a slight increase in the cash cost of production, it may take several days to isolate the cause and take remedial action as a result of the delay in assembling and compile information from disparate data sources. In the mean time, the mine will be operating sub-optimally.

The existence of specialized applications, each using distinct data base structures and query routines, makes IT support more labour intensive and costly than is necessary. In addition, the tools become less flexible, since specialized knowledge is necessary in order to customize them. In the face of declining sales, maintenance, modification and training contracts have become a major source of income for many mining software suppliers. These act as a disincentive in addressing the issues outlined in this paper.

3 BRIDGING ISLANDS OF DATA

It comes as no surprise that the problem of disparate databases first manifested itself in larger mining companies. Many of these companies are currently undertaking initiatives to bridge the data islands which exist within their enterprises. These initiatives can be grouped into three categories:
- data warehousing,
- mine management systems, and
- company-wide data standards.

3.1 *Data warehousing*

A data warehouse is a separate platform with its own server(s), relational database management system, GUI tools, query and reporting tools, data mining, hygiene and archaeology tools. It is a decision support tool that allows executives to take 'near-time' decisions. Figure 1 illustrates the basic concept of a data warehouse. The warehouse acts as a separate layer in the information architecture, and permits one way flow of selected data from specialized applications such as the mine planning system through specially written interfaces. This data is stored in the system's relational database for processing and retrieval on demand. The depositing of data into the data warehouse is automatically managed by each application. In some cases, transmits can be every minute, but in others, transmits may be weekly (Goddard & Tremblay, 1997).

A criticism of data warehousing is that it necessitates an additional application to resolve the problems of integrating existing applications. An extra level of complex-

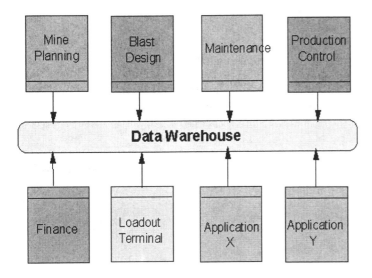

Figure 1. Data Warehouse Concept (adapted from Goddard & Tremblay 1997).

ity is added to the existing IT architecture which will have the effect of increasing, not decreasing the level of software support necessary to maintain applications in the mine.

3.2 *Mine management systems*

An alternative solution is to replace many of the existing applications with an integrated mine management system. These are usually large client-server based systems manufactured by one company. In the past, such systems were specific to mining, but more recently a number of systems are being installed which have enjoyed popularity in the manufacturing and process industries. As an example of the latter, Codelco Chile is currently installing the R/3 System manufactured by the German company SAP in all of its mining and downstream processing operations. This system will replace existing financial, accounting, maintenance, materials management, human resources, project management and quality control applications in all divisions (Codelco-Chile, 1997). However, having been developed for the manufacturing and process industries, several legacy software applications specific to mining will have to be maintained and, if possible, integrated into the SAP system. These include geological modelling, mine planning and production control or dispatching systems. Like many other activities in the mining industry, installation, management and maintenance of the SAP system has been outsourced.

Such solutions represent somewhat radical and ambitious solutions to the problem of integrating disparate data sources. They also represent considerable capital investment and ongoing maintenance expenses. However, the end result will be uniform data structures and querying languages, common reporting standards and the ability to deliver near-time information.

3.3 *Company-wide standards*

A third, and less costly solution to the problem of application integration is to stipulate company-wide data storage and retrieval standards. Such policies attempt to influence purchasing or development decisions to maintain commonality between applications. Because many mine planning applications use sequential data structures, in practice it can sometime be difficult to adhere to such standards. One mine which has very successfully implemented a standardized approach to data management is the El Teniente Mine in central Chile. Here, many specialized applications have been developed in-house. All use Oracle databases and SQL calls for data retrieval. As a result of this consistency, it is a relatively easy task to develop management decision support tools capable of accessing relevant information within the departmental databases.

4 MINING INDUSTRY INITIATIVES

Specialized mining software suppliers appear to be doing little to address the problem of interoperability. An exception to this rule is Lynx Geosystems, who, in association with Integrated Software Systems, have approached the problem of integrating specialized vendor software with general vendor software through the addition of a third component known as 'the Miner's Workbench' (Rychkun & Higgens, 1996). The Miner's Workbench is an application which functions as the glue to bind the first two components together. The concept is similar to that of a data warehouse, but with additional functionality to support CAD and GIS systems as well as a common database support.

In Australia, a consortium of eight mining companies, through AMIRA (The Australian Minerals Industry Research Association) have been active in supporting the development of a Data Model and Data Translator (Henley, Hornby & Smith, 1996). The Data Model, developed by CSIRO (Commonwealth Scientific and Industrial Research Organization), is based on geometric objects and is suitable for handling geoscience and mining data of many different types. The Data Model provides a general purpose geological data translator to ease the problem of data transfer amongst many different proprietary packages. The Data Model is an 'open' development in the sense that CSIRO intends to make it available to bone fide applications developers as a framework for future products (Henley, 1996).

It is interesting to note the current efforts of some of the major equipment suppliers to integrate equipment and mining applications into complete mine management systems. In April 1997, Caterpillar Inc. signed a letter of intent to purchase a minority interest in Brisbane based Mincom, best known as a supplier of integrated mine management systems. The company also acquired a controlling interest in Aquila Mining Systems Ltd., a Montreal based company specializing in GPS monitoring equipment for shovels and drills. In addition, Caterpillar have developed a Computer Aided Earthmoving System (CAES) which enables daily mine plans to be downloaded and displayed in mobile equipment via radio modem links (Greene, 1997). The heart of the CAES system is the software package known as the 'Mining and Earthmoving Technology manager'. METS manager provides a data translator (Figure 2) which

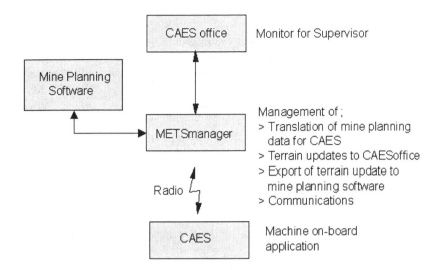

Figure 2. Architecture of Caterpillar CAES System (Greene, 1997)

enables mine plans created by third party products to be translated to CAES format and broadcast to mobile equipment. Updated topography resulting from GPS monitoring of mining activities can be exported back to the mine planning software to close the loop between planning, production and reconciliation.

Caterpillar are not alone in their endeavours to develop and supply integrated mine management systems. Also in 1996, Komatsu Mining Systems Inc. bought a controlling interest in Modular Mining Systems, a major supplier of open pit and underground production control and reporting software. In the 'intelligent mine' of the future, it is expected that Modular's Dispatch™ system will form the interface between production equipment and applications such as mine planning, maintenance, and accounting systems (MMS, 1997).

5 OPEN SYSTEMS STANDARDS

So why are certain companies and research organizations investing large amounts of time and money in developing data translators, installing extra software layers to bring interoperability to applications, or replacing existing applications with management systems that promise greater interoperability and compatibility? Might not a feasible alternative be for interested industry representatives and software suppliers to form a consortium dedicated to examining the issue of mining industry software standards? Two examples of such consortia in related industries are: POSC, the Petroleum Industry Open Systems Corporation, and MIMOSA, the Machinery Information Management Open System Alliance. Both of these initiatives will be discussed in more detail.

5.1 *POSC*

Petroleum companies typically become involved in joint ventures for exploration and production in order to minimize risk and allay costs. Typically, collaborative effort is necessary amongst geologists, geophysicists and other groups. However, information sharing amongst the specialist companies was found to be extremely difficult due to the multitude of separate and incompatible systems, tools and data representations of data used by each company. To address this issue, in late 1990 the *Petroleum Industry Open Systems Corporation*, POSC, was founded by BP Exploration, Chevron, Elf Aquitaine, Mobil Oil and Texaco. Since then, its membership has increased to include other petroleum companies and their software and hardware vendors (Devine & Brogden, 1994).

The POSC founders envisioned an open platform that could integrate functions and data across multiple applications running on various computers in a distributed environment. All components of the overall architecture were considered: data and data models, technical applications, software integration platforms, user interfaces and hardware. A set of standards specifications were developed, which included (Devine & Brogden, 1994):

- POSC Software Integration Platform Specification – Base Computer Standards
- POSC Software Integration Platform Specification – Data Access and Exchange
- POSC Software Integration Platform Specification – Exchange Format
- POSC Software Integration Platform Specification – Exploration and Production User Interface Style Guide
- POSC Epicentre Data Model (Volumes 1, 2, & 3)

The ability of POSC to influence vendor design was confirmed when ORACLE and UNI/SQL announced that they would form an alliance to ensure that their products were POSC compliant (Whiting, 1994).

5.2 *MIMOSA*

The *Machinery Information Management Open Systems Alliance* was formed in September 1994 by a group of suppliers of machine condition monitoring instruments, end-user and service companies (Bever, 1997). The alliance was established in response to:

- the lack of timely access to machinery information,
- the need to integrate islands of data, and
- the need to provide end-users with integrated decision support tools to support enterprise asset management.

Since it was founded, MIMOSA has grown to include 60 participating companies. This includes 12 of the 16 leading suppliers of condition monitoring tools. It was felt that the development of a set of open standards specifications would form the basis for the design of next generation plug-and-play systems, would save end-users from the burden of ongoing integration efforts and would provide end-users more freedom to select the best technology available. A key aspect of MIMOSA's charter is to:

- adopt open conventions for standard data formats, and
- recommend protocols for seamless information exchange.

Their progress to date includes:

- the approval of the MIMOSA Common Relational Information Schema (CRIS) specification version 1.0,
- an approved MIMOSA export data (MED) file specification to allow batch import and exports,
- development of a database storage standard for machine vibration data, and
- a draft version of MIMOSA programmable database interface which defines MIMOSA SQL/CLI functions.

Many of the documents developed by MIMOSA are available via Internet from: *http://www.hsb.com/mimosa*

5.3 *A mining industry open systems alliance?*

The above initiatives address many problems which are similar in nature to those currently being experienced by the mining industry. Indeed, the collaborative nature of petroleum exploration work that lead to the formation of POSC mirrors the business practices of mining companies today. No longer is a single company solely responsible for exploring, developing and exploiting a deposit. In the production process alone, it is now common for work to be outsourced to a multitude of contractors with responsibilities for either drilling, blasting, or hauling broken rock or for providing services such as catering, IT support and engineering. The presence of many companies working in collaboration without adequately defined data management and exchange standards is a recipe for cost overruns associated with the inefficiencies of accessing and exchanging information.

6 CONCLUSIONS

As suggested by the title, this paper suggests the formation of a Mining Industry Open Systems Alliance, or Corporation (MOSA, or MOSC). In the same manner as the MIMOSA initiative, the principal focus of a Mining Industry Open Systems initiative would be to:

(i)   adopt open conventions for standard data formats, and
(ii)   recommend protocols for seamless information exchange.

By adopting such an initiative, mining companies could expect superior compatibility and interoperability of applications software. This in turn will reduce IT support costs, enable access to near-time information for decision support, and permit companies the freedom to select "best technology".

The main barrier to the adoption of such a proposal is likely to be resistance amongst suppliers of specialized mining software. Suppliers may be reluctant to make their products more accessible for fear of losing hard won existing market share, and for fear of losing the maintenance, modification and training contracts that go hand in hand with their products and, in some cases, underpin their survival. However, in a MOSC, (or MOSA) compliant environment, leading mining software suppliers will likely benefit by increasing their market share as a result of "captive sites" converting to open systems standards.

REFERENCES

Bever, K., 1997. *MIMOSA: Bridging Islands of Technology.* Presentation notes.

Devine, M. & Brogden, I., 1994. *Information Technology Planning: Critical for Implementing Advanced Manufacturing Automation.* Proc. 6th Canadian Symposium on Mining Automation, Montreal, Oct., pp.239-256.

Goddard, G.J. & Tremblay, P., 1997. *Mining Information Using a Data Warehouse.* CIM Annual General Meeting, Vancouver, April.

Codelco-Chile, 1997. *Proyecto Swing - Un Lenguaje Comon.* Codelco Horizontes, April/June.

Greene, D., 1997. *Linking Mine Planning to Operations via Computer-Aided Earthmoving Systems.* CIM Annual General Meeting, Vancouver, April.

Henley, S., 1996. *Mining Software - A Hard Landing.* Mining Magazine, May, pp.316-319.

Henley, S., Hornby, P. & Smith, K., 1996. *Intelligent Data Exchange Based on a Consistent Data Model.* Proc. APCOM XXVI, Pennsylvania State University, Oct., pp.521-522.

Modular Mining Systems Chile S.A., 1997. Personal Communication. Nov.

Rychkun, E. & Higgins, D., 1996. *The Miner's Workbench: A New Solution to an Age Old Problem.* Earth Science Computer Applications, Vol.12, No.s2 & 3, Oct & Nov.

Whiting, R., 1994. *Alliances will close OO, Relational Gap.* Client/Server Today, Aug.

GEOSTATISTICS: A CRITICAL REVIEW

Stephen Henley
Resources Computing International Ltd / Snowden Associates (Europe), London, UK

David F. Watson
CSIRO Exploration and Mining, Perth, Western Australia

ABSTRACT: Geostatistics has become the accepted standard model for mineral resource estimation. An examination of the underlying assumptions of geostatistics is followed by a review of some theoretical and practical problems in the use of linear and nonlinear geostatistics. Possible alternatives to the geostatistical model are examined, and some less restrictive approaches are presented.

KEYWORDS: geostatistics, resource estimation, evaluation, natural-neighbour, nonparametric statistics, geology

1 INTRODUCTION

Over the past 20 years, geostatistics has become the internationally accepted standard model for mineral resource estimation, and has found application also in a number of other fields. This paper will concentrate on its suitability for mineral resource applications, as this is the purpose for which geostatistics was primarily developed. It is not an attempt to comprehensively discredit geostatistics as in recent works by Shurtz (1997) but rather to point out some ways in which the geostatistics model might not be appropriate, and some internal inconsistencies in certain variants of the geostatistical model, which should require great caution in the application of these techniques. We show some possible different approaches which avoid the difficulties and inconsistencies inherent in geostatistics.

Geostatistics as a body of theory and applications had its origins in the South African gold mining industry, where Krige (1951) developed moving-average methods for providing better grade estimates than the previous standard polygonal technique. A theoretical mathematical framework was subsequently developed by Matheron (1962) and co-workers in France. Linear geostatistics was found to have a number of very convenient numerical properties which led to its acceptance by the mining industry during the 1970s and 1980s. However, it was found that as a model its assumptions were too restrictive, and so progressively a large number of modifications have been developed in attempts to overcome these shortcomings. Even linear geostatistics exists in different 'flavours' – simple kriging, ordinary kriging – depending on whether or not the user considers that a global mean value is 'known'.

One of the problems in using geostatistics nowadays is that it has largely become the preserve of an elite group who understand the terminology and the theory behind it. In some of the more modern variants, that theory is becoming ever more impenetrable to non-mathematicians, and there is increasing uneasiness within the mining industry over the use of such techniques as a 'black box' to provide estimates of quantities which are at the very basis of the industry's performance – the grade and tonnage of ores. Although it was not geostatistical estimation which was at the root of the Bre-X affair in 1997, it easily could have been, and it is probably only a matter of time before some similar scandal erupts over the numerical methods used for resource estimation.

In view of this, there have been a number of developments away from the increasingly obscure mathematical theory towards more practical – and more easily understood – methods. The most widely applied of these was developed by Journel and co-workers at Stanford, and is known as 'indicator kriging'. This was developed as a very simple practical solution to one of the more intractable problems in geostatistics, how to deal with local probability distributions which are not Gaussian.

There have also been attempts to work from basic principles, to re-define the problem. These have included approaches based on natural-neighbour interpolation (Sibson, 1981), on the use of nonparametric statistics (Henley, 1981), and others. Serious questions on the entire theoretical basis of geostatistics have been raised by Shurtz (1997) which in the authors' opinion have not satisfactorily been answered – and unfortunately Shurtz himself offers no suggestions for satisfactory alternative methods. However, such studies and techniques, developed in isolation by individuals without the support of an established research group or the mining industry itself, have failed to make much impact on the practitioners of geostatistics, and have been virtually ignored by the industry.

This present paper attempts to identify the key issues which must be addressed in finding a valid method of mineral resource estimation. It starts without any preconceptions and without any presumption that geostatistics – as developed over the past 40 years – is the best approach, or even provides a valid set of methods to be used.

It is anticipated that there could be a strong reaction from the geostatistical community, with suggestions that the authors do not understand geostatistics, or are exaggerating the problems, or are just plain wrong. To answer these allegations in advance, both authors have worked in the field of mathematical geology for very many years. The first author is a geologist who has been involved in this field since 1967, has used geostatistics since 1971, and has substantial consultancy experience of applying geostatistics to real mining and exploration situations since 1978. His concerns with the validity of these methods began early, and were voiced originally in a book published in 1981 (Henley, 1981). The second author is a mathematical geologist with a background in geological exploration dating from the early sixties. He has published a book (Watson, 1992) surveying computer methods for resource evaluation.

2 THE NATURE OF GEOLOGICAL SAMPLE DATA

Geological materials ('rocks' in the most general sense) at the smallest scales are heterogeneous collections of mineral grains, interstitial fluids, and solid and fluid inclusions. At larger scales, there are other heterogeneities – bedding planes, mineral

12

banding, veinlets, concretions, and so on at the macro scale, which may further be complicated by folding and fracturing. The concept of 'homogeneity' at any geological scale is therefore, at best, only a convenient simplification.

For the purposes of resource estimation, a 'point' is in fact a sample consisting of a section of core, or a collection of cuttings, or some other volume of rock, which is asserted to be 'typical' or 'representative', in some sense, of a larger volume. A sample is normally homogenised by crushing, grinding, and thorough mixing before analysis (the difficulty and expense of such a process is surely in itself evidence that rocks are not homogeneous even on a small scale?). The sample size is recognised by geostatisticians as the 'support' of the values from that sample. Such a sample contains a true quantity of each chemical element, which in principle can be analysed as accurately as desired. In practice of course there is always some significant sampling error, which consists of a combination of the error due to incomplete homogenisation and the analytical error. In most geostatistical studies, the magnitude of such errors is completely ignored, though Gy's sampling theory makes it quite clear that within-sample error is significant and should be carried through any further analysis. A sample is drawn at a position in space from a real geological population of all possible samples spread through the three-dimensional region of interest. This real population is not random but is defined by the set of usually complex physico-chemical processes which have created it.

Every sample is unique and non-reproducible. Because of the averaging effect of taking a macroscopic sample composed of an arbitrarily large number of microscopic components, we can say that the population of all possible such samples may have spatial continuity at the scale of the sample. Even though it is not physically possible to collect overlapping samples, nevertheless in principle there is continuity between the true chemical concentrations of two adjacent samples. Even in the presence of a nugget, this continuity exists because any proportion of the nugget may be included in a sample, simply by moving the sample boundary. A 'nugget effect' in fact represents not a discontinuity but merely a smaller scale of continuous variation.

So far in the discussion, the only variable which is subject to a statistical treatment is the analysed chemical concentrations of some element in a sample:

$$x = x_t + e$$

where
x is the observed and reported concentration
x_t is the unknown true concentration
e is the unknown error term – the combination of all sample and analytical errors

At the macro scale upwards, geology can and does exhibit local zones of rapid variation (such as bedding or intrusive rock boundaries) or real discontinuities (such as faults), which may divide the region of interest into separate domains which are themselves often considered to be homogeneous. However, as has been remarked already, there is no real homogeneity at any scale: this is just a convenient fiction.

So what sort of beast are we dealing with statistically? Is there such a thing as a 'random field' or a 'regionalised variable' in geology? The answer is quite clearly that the real population can be highly complex but is certainly not random. It must

13

always be remembered that random fields and regionalised variables are merely models which are applied to the data and should not be elevated beyond that status. Their usefulness is limited to the quality of their performance in modelling the real world.

Thus the entire body of geostatistical theory which has been developed should be judged by two criteria:

(a) its internal self-consistency

(b) its performance in describing the 'real world'.

We concentrate in this paper on the first of these, because if geostatistical methods are not even self-consistent – i.e. they violate their own axioms – then even quite good performance most of the time should not be sufficient to justify their continued use.

3 PROBLEMS WITH GEOSTATISTICS

Regionalised variable theory is based on a number of concepts, some familiar from classical statistics and others which would appear quite strange to a statistician.

The principal assumptions – the axioms of regionalised variable theory as defined by Matheron – are as follows:

- stationarity (of various types) of the variable to be estimated
- homoscedasticity: variance does not change with location
- continuity of data distributions
- additivity of the variable

The estimation procedure itself (known as kriging) is a special case of a weighted moving average in which the weighting factors are derived from covariance relationships among the sample locations and the point or block to be estimated.

3.1 *Kriging performance*

Let us at the beginning accept that in some particular situation all of these assumptions hold. Shurtz (1996) points out a number of examples where the use of standard linear geostatistics yields unreasonable and highly improbable estimates. Some of his examples themselves present highly improbable sampling geometries. However, one simple example which he quotes is examined here, as it is of particular importance in resource modelling because of its resemblance to the geometry of real drillhole data. This is the 'Linear Pattern', which in Shurtz' example has a set of 13 data points ranging from locations $X = 0, Y = -3.0$ to $X = 0, Y = +3.0$ in steps of 0.5 units. In estimating the value at $X = 7, Y = 0$ Shurtz shows that the weighting factors allocated to the two points at the ends of the line of samples (samples no. 1 and 13) are an order of magnitude larger than those on any of the other sample points. Shurtz' example has been reproduced using an industry standard kriging program within the *Datamine* mining software system, as indicated in Table 1.

Of course, the linear semivariogram may be convenient for demonstrating a case, but it is rarely used in real resource estimation projects. The standard semivariogram model used is the spherical model. The same data set has been used with a spherical model semivariogram ($c0 = 0$, $c = 1$, $a = 20$) to resemble more closely a real situation (Table 2).

Table 1. Linear pattern using linear semivariogram.

NO. OF SUBCELLS = 1 1 1
SUBCELL DIMENSIONS = 0.10 0.10 0.10
VARIANCE OF A SAMPLE IN SUBCELL = 0.00

SampNo.	ID	X	Y	Z	Dist	Value	Covariance	Kriging Weight
1	1	0.0	3.0	0.0	7.6	1.000	0.381	0.317
2	2	0.0	2.5	0.0	7.4	0.000	0.372	0.030
3	3	0.0	2.0	0.0	7.3	0.000	0.364	0.032
4	4	0.0	1.5	0.0	7.2	0.000	0.358	0.033
5	5	0.0	1.0	0.0	7.1	0.000	0.354	0.035
6	6	0.0	0.5	0.0	7.0	0.000	0.351	0.035
7	7	0.0	0.0	0.0	7.0	0.000	0.350	0.036
8	8	0.0	-0.5	0.0	7.0	0.000	0.351	0.035
9	9	0.0	-1.0	0.0	7.1	0.000	0.354	0.035
10	10	0.0	-1.5	0.0	7.2	0.000	0.358	0.033
11	11	0.0	-2.0	0.0	7.3	0.000	0.364	0.032
12	12	0.0	-2.5	0.0	7.4	0.000	0.372	0.030
13	13	0.0	-3.0	0.0	7.6	1.000	0.381	0.317

LAGRANGE MULTIPLIER - 1.000 0.231
PANEL AT 7.00 0.00 0.00 VALUES 0.63 0.60 13

Table 2. Linear pattern using spherical semivariogram.

NO. OF SUBCELLS = 1 1 1
SUBCELL DIMENSIONS = 0.10 0.10 0.10
VARIANCE OF A SAMPLE IN SUBCELL = 0.00

SampNo.	ID	X	Y	Z	Dist	Value	Covariance	Kriging Weight
1	1	0.0	3.0	0.0	7.6	1.000	0.544	0.325
2	2	0.0	2.5	0.0	7.4	0.000	0.532	0.028
3	3	0.0	2.0	0.0	7.3	0.000	0.522	0.030
4	4	0.0	1.5	0.0	7.2	0.000	0.514	0.032
5	5	0.0	1.0	0.0	7.1	0.000	0.508	0.033
6	6	0.0	0.5	0.0	7.0	0.000	0.505	0.034
7	7	0.0	0.0	0.0	7.0	0.000	0.504	0.034
8	8	0.0	-0.5	0.0	7.0	0.000	0.505	0.034
9	9	0.0	-1.0	0.0	7.1	0.000	0.508	0.033
10	10	0.0	-1.5	0.0	7.2	0.000	0.514	0.032
11	11	0.0	-2.0	0.0	7.3	0.000	0.522	0.030
12	12	0.0	-2.5	0.0	7.4	0.000	0.532	0.028
13	13	0.0	-3.0	0.0	7.6	1.000	0.544	0.325

LAGRANGE MULTIPLIER - 1.000 0.324
PANEL AT 7.00 0.00 0.00 VALUES 0.65 0.86 13

It will be noted that the effect is indeed even stronger in this case, with even higher weightings (at 0.325) on the end two sample points. All of the intermediate samples are geometrically closer to the point being estimated, yet have extremely low kriging weights.

What this means in practice is that given a common situation, of a drillhole ore intersection sampled from just above to just below the ore zone, estimation of grades at short distances to either side will virtually ignore any high grade samples in the middle and will compute an estimate from only the outermost pair of samples. This applies in principle to every variety of kriging which has been developed – with one or two exceptions which are notorious rather than notable. Recognising this problem, Deutsch (1994) proposed a number of workarounds. Unfortunately these all involve artificially changing the data configuration – in other words, massaging the data in ways which falsify the true sample relationships and are therefore invalid. Shurtz has demonstrated also that these workarounds lead to inconsistent results and are therefore doubly unacceptable.

Shurtz documents a number of other situations – some realistic, others more artificial – in which it is clear that the normal application of standard geostatistical methods leads to results which are plainly inconsistent, unstable, and unlikely.

However, the situation becomes even worse if we examine the validity of the assumptions which are made in using geostatistics. It should be borne in mind that all of these assumptions relate just to the geostatistical model: few geostatisticians would attempt to claim that these are in fact real properties of geological data. Geostatistics stands or falls by the appropriateness of the model. If the properties of the model are too divergent from those of the real world which the model purports to represent, then the usefulness of the model must be called into question.

3.2 *Stationarity: the intrinsic hypothesis (zero'th order stationarity), and first order stationarity.*

There are three types of stationarity which commonly have been required by different geostatistical methods. These are discussed in David (1977, p.92ff).

Second-order stationarity, also known as the weak-stationarity assumption, requires that the expected value is the same everywhere, and that the spatial covariance function is also the same everywhere.

The intrinsic assumption, which is a weaker form of stationarity, does not impose a constant expected value, but does require that the expected value of $[Z(x) - Z(x+h)]$ be zero for all vectors h separating two points in the region of interest. The variance of this increment $VAR[Z(x) - Z(x+h)] = 2 (h)$, and defines the variogram (hence (h) is the semivariogram).

In the method known as universal kriging, a weaker form of stationarity is allowed, in which the expected values of both $Z(x)$ and $[Z(x) - Z(x+h)]$ may vary regularly with location. In universal kriging, this regular variation is modelled by a polynomial function. There are also still weaker stationarity assumptions which can be used, and have been identified in the theory of generalised random intrinsic functions. However, all of linear model geostatistics requires at least intrinsic stationarity.

The geostatistical models which have been developed to use weaker stationarity assumptions are all much more difficult to use and more difficult to understand, and also – like universal kriging – have methodological problems such as the difficulty of obtaining a semivariogram model.

Without a semivariogram – which requires the intrinsic form of stationarity to hold (or a simple transformation to it, such as universal kriging purports to offer) – there is no geostatistics.

3.3 *Homoscedasticity: variance does not change with location*

In many geological situations of interest to the mining industry this is quite obviously untrue. For example there can be extreme local variability of parameters depending on the location in the middle, on the margins, or outside an orebody. Geostatisticians have tried to overcome this by assuming that there is a 'proportional effect' (local variance is proportion to local mean value), but this holds true only for lognormal distributions, which it has been established are usually inappropriate – indeed this is why lognormal kriging is so little used. The 'proportional' effect is usually true only in a very qualitative sense. Another way in which geostatisticians have tried to avoid this problem is by definition of ever smaller 'homogeneous' zones. This, of course, often leads them into the trap of identifying zones which contain too little data from which any useful statistics may be derived.

3.4 *Continuity of data distributions*

This is something which again is too often taken for granted. There are many arguments about whether data should or should not be required to follow a normal (Gaussian) distribution, but all are agreed that the data must at least follow a continuous distribution – in other words must be measured on an interval scale. Nevertheless, one of the most widely used geostatistical methods – indicator kriging – deliberately violates this condition in the most drastic way possible, by using data which are defined as a discrete distribution containing only 0s and 1s. Proponents of indicator kriging would claim that these 0s and 1s represent merely particular samples of what is 'really' a continuous distribution – and that intermediate values obtained by kriging demonstrate such an underlying continuous distribution. However, this argument is patently false, as by definition all data points are assigned values of 0 or 1, and thus occupy maxima or minima on the distribution which even the method's supporters must admit is restricted to this range and thus very different from the gaussian or other unbounded distributions which were envisaged when geostatistical theory was developed.

3.5 *Additivity of the variable*

This assumption is almost invariably ignored by geostatisticians. Nevertheless it is *absolutely* required by the linear model on which all of geostatistics is based. Linear kriging is a special case of weighted averaging. Averages cannot legitimately be computed on data for which the addition operation has no meaning. Therefore, certain commonly used geological variables which are not additive cannot be used for kriging. These include such values as permeability – though porosity, recorded as a linear function of pore volume per unit volume, is additive. Also, *all* ratios in which the denominator value varies are non-additive. These include such ratios as Zn/Cu. *All* measurements recorded in or transformed to logarithmic or other nonlinear scales are also non-additive. These include such values as pH and the grain-size measure phi.

   In indicator kriging, the indicator values themselves are not in any sense additive – indeed they are not even recorded on the real or integer measurement scales in which addition is defined as an operation. In probability kriging, the probability values which are being kriged are also non-additive. And further, in any version of kriging where nonlinear transforms are applied to the data – such as disjunctive and multi-gaussian kriging, the resulting variables are non-additive and therefore the result of transforming data to a 'better behaved' distribution is to invalidate the method used for estimation of values from it.

Normally, ordinary metal grade values (expressed as grams per tonne or weight percent) are assumed to be additive. However, in base metal or industrial mineral deposits, the grade frequently has a significant effect on the density. In such cases, when modelling blocks which are defined, geometrically, by volume rather than tonnage, the grade value itself cannot be considered as additive. Adjustments must be made for varying density, and the variable to be modelled in such cases must be computed as grams or kilograms per cubic metre for example. Where the deposit is tabular in form, an alternative approach is to model metal accumulations, or grade multiplied by thickness (and by density). This reflects the traditional practice in the South African reef gold mines where grade was formerly recorded as inch-pennyweight.

It is seen that this requirement of additivity is actually rather strict, and that its violation – which has become very common – invalidates a very wide range of geostatistical methods. Some methods such as indicator kriging, which have become widely used, are invalid from the outset because of their violation of this most basic and important assumption.

3.6 Geostatistical Practice

In view of the frequent violation of one or more of the basic assumptions behind geostatistics, it might be asked why such a variety of methods have become so entrenched in 'standard practice'. The answer is perhaps that enthusiasts have become so carried away by the elegance of the mathematics that they have lost sight of the need to ensure its validity.

Apologists when confronted by the more glaring problems generally use the excuse that the linear model in geostatistics is actually quite robust to moderate deviations from the assumptions. Be that as it may, the result is all too often that geostatistical methods – including the more advanced nonlinear methods – are applied without a thought to the real data properties. This can and does produce situations in which forecasts of mineral grade and tonnage fail to be realised in production. Partly this can be explained by natural properties of the linear block estimation method itself, and it is possible that sometimes corrections can be applied. However, more often than not these corrections themselves are found to be inadequate (or just wrong). *The fault lies in the choice of an invalid estimation method.*

Another excuse often used is that whatever geostatistical method is being used is merely a 'model' of the real world, and not intended to be a precise representation of it. However, the quality of any linear, additive, model of a quantity such as pH or permeability must be open to serious question. The quality of an orebody model which assumes stationarity of grades (or even of the semivariogram) over a gold orebody containing very high-grade patches must also be questioned. *An unqualified assertion of robustness is a very weak statement to support a family of numerical techniques which the mining industry is expected to use as the 'best' way to estimate its asset base.*

In the face of proof that assumptions, whether of stationarity, additivity, or continuity, are violated, assertions that the geostatistical model is robust to the frequent violations of its own axioms are a weak response to the obvious demand that a more appropriate and valid model must be found.

4 THE USE OF STATISTICAL MODELS

Statistics has become an essential part of scientific studies. In the experimental sciences, such as 'pure' chemistry, physics, and some of the biological sciences, and also in engineering, it is possible to set up experiments which are reproducible to as close a degree as required. In analysing the results of such reproducible experiments, classical parametric or nonparametric statistical methods are clearly suitable techniques to use. They were designed for this purpose.

In contrast, in the observational sciences such as geology, meteorology, and astronomy, situations and events tend not to be reproducible. The result is that there is usually only one realisation of any 'experiment'. It is not possible to apply normal statistical methods in such cases. Geostatistics is an attempt to establish reproducibility - and a sample size greater than one – by using its rather strict stationarity assumptions. There are certainly experiments in geology to which statistics can validly be applied, such as the analysis of grain size in a sample of sand or gravel, or biometrics in a sample of fossils of a given species.

However, when the objective is characterisation of a volume of rock which is known to be inhomogeneous – and in ore deposits the inhomogeneities, and the anomalous values, are usually not only known but deliberately sought – it is clearly inappropriate to apply a model whose basic assumptions require spatial stationarity. Poorly substantiated assertions of inherent 'robustness' of the geostatistical model to nonstationarity cannot hide the fact that the choice of the model itself is unsuitable.

5 A NEW PARADIGM?

In the estimation of unknown grades at points (and by extension, average grades in blocks) among irregularly distributed data points, there are several questions which need to be answered. These include:
- what type of method (or algorithm, or formula) is the 'best' for estimation of the value at a point?
- how should data points be weighted to reflect their relative distances from the point to be estimated?
- how can the data points best be weighted to reflect and compensate for the inevitable 'clustering' effects of irregular sampling

Geostatistics, as it is normally applied, posits the linear model and not surprisingly advocates the use of a simple weighted average. In the multiple nonlinear methods which have been derived from the original linear model, even the limited validity which in its basic form it does possess is lost, and additional assumptions are required. In the extreme case of indicator kriging, the most basic assumption of regionalised variable theory – that the distribution at least be continuous – is itself violated.

Even though the practitioners themselves know that the underlying distribution is continuous, they nevertheless set up a series of discrete 'indicator' distributions containing values of only 0 or 1 indicating whether the data value at each point is above or below a particular cutoff. The application of geostatistical methods to this discrete distribution naturally produces a distribution of estimates in which values normally lie within the range 0 to 1. The statistical properties of the distributions of these estimates are of course totally different from those of the indicator samples.

It is clear that the linear model itself is the problem. It is too constrictive a framework within which to develop a valid and viable set of methods for estimating values in real geological populations which are neither stationary nor distributed according to the Gaussian or any other simple statistical law.

If statistics can provide an answer, there are some statistical models which are much less demanding of the data than the linear model, and these are known collectively as 'nonparametric statistics'. Perhaps some of these may form the basis for a more robust and more valid estimation model for geological data. In view of the fact that many geological data – especially those of interest to the mining community – are highly skewed, often with long tails of extreme high values, it would seem that some method which is robust to large deviations from the normal distribution would give a better starting point.

5.1 *Estimation of the value at a point*

First, it must be decided what type of model should be adopted. At one extreme it can be argued that there is complete spatial continuity, and that the value at any point is determined entirely by the values at immediately adjacent points as on a rubber sheet. In this scenario, there is no statistical component, and the problem is merely one of geometry. However, since the geometry itself is incompletely known, some statistical assistance is required in order to obtain the point estimate (in geostatistical terms, this is the situation of zero nugget effect with continuity at the origin as in a gaussian semivariogram).

At the other extreme, it can be argued that each point is independent of its neighbours and can be determined solely by statistical estimation based on some assumed distribution (in geostatistical terms, this is the situation of pure 'nugget effect').

In practice, a combination of the two is clearly the most appropriate. Both extremes are unrealistic in most situations. When modelling geological structure, the surface is commonly deterministic to a close approximation – at least in simple deposits such as coal seams – but there are almost always some small-scale irregularities which may not be predictable. When modelling a gold deposit, the locations of grains which control the grade may appear to be random on all scales, but there is generally some more regular underlying trend which assists in prediction of grades.

The question that must be asked now is whether there is – even theoretically – any single estimation method which can give a 'best' estimate in both these cases and in every situation in between?

The answer is probably 'no'. In the case of geological structure, the best type of estimation model is surely one which models the physics of rock deformation, conditioned by the known data points. A simple version of this might be a spline model or a Briggs minimum curvature model. A more complex model which is geologically more realistic might be obtained from the use of numerical finite element or finite difference methods which model the physics of the rock units themselves. However, given the complexity of such solutions – and the requirement to supply realistic physical parameters and some knowledge of the geological history – such methods are usually impractical.

In other cases, especially for grade estimation, quite clearly the 'rubber-sheet' model is insufficient, and some statistical model must be used. Since many grade variables are minor or trace components, the distribution of their grade tends to be positively skewed - in contrast with major rock-forming components such as silica or alumina which tend to be normally distributed or even negatively skewed.

Henley (1981) proposed a range of statistical methods for estimation of point values from surrounding data points. These methods are based on nonparametric estimates such as the median, which are very much more robust than the mean to asymmetric or other ill-behaved distributions. In particular, use of the median can correct automatically for bias due to tails of extreme values. This method is suitable for use in the 'pure nugget-effect case' where there is no evidence for spatial continuity.

Usually, of course, there is a mixture of continuity and randomness. We have a 'fuzzy rubber sheet'. In such cases, we need to take into account the form of the surface surrounding the point to be estimated. This could be (and has been) done by using such devices as local least-squares polynomial fitting. However, such methods are notoriously unstable in the presence of non-ideal data with spatial clustering or extreme grade values. A much better approach – in terms of its stability – would be a method which uses rank-order statistics rather than absolute values. The question of spatial clustering effects is dealt with separately, below. What we require is an indication of local surface form at the point to be estimated – as first and second partial derivatives. One way to do this is to use the Kendall rank correlation coefficient between grade and distance away from the point in different directions. Bear in mind that we are seeking only qualitative surface form information. For this reason, the precise definition of a search radius is much less important than in the case of linear-model geostatistics.

If there is high positive correlation in all directions away from the point, this is an indication that the point represents a trough: the median is thus an unsuitable measure and a low quantile would be more appropriate. Conversely if there is high negative correlation away from the point, it lies at a peak and a high quantile should be used. Intermediate cases would be estimated better by other quantiles. In the case where there are an equal number of positive and negative correlations away from the point, then the median is an appropriate estimate.

5.2 Distance weighting

Linear model geostatistics uses the semivariogram to determine weightings. This does provide, in general terms, a weighting which decreases with distance (strictly speaking, decreases with autocorrelation) but is also dependent on the spatial relationships among the data points. Classical moving average estimation methods usually use a much simpler weighting function which is related purely to the distance – for example inverse distance squared – with a cutoff at a defined search radius. However, they do not make any allowance for clustering effects.

There is a problem which is common to any simple distance weighting approach and to the weightings used in kriging. This problem is that weightings are not related to the value at an observation point. It is intuitively obvious that in, for example, a gold deposit, a low-grade sample will be much more representative of a large volume around it than a high-grade sample. This is the true nugget effect. Yet all geostatistical methods are blind to this truth. They allow exactly the same weightings – controlled by the semivariogram function – to be applied whatever the grade of each sample. This leads to the commonly found nonsense of large high-grade patches in otherwise sparsely sampled parts of an orebody. This question is discussed more fully in Section 7 below.

5.3 *Declustering*

One of the more sweeping claims made for geostatistics is that it automatically compensates for clustered data. As Shurtz has conclusively shown, and as has been demonstrated in Section 3.1 above, this is patently false. Linear-model geostatistics certainly does allot different weights to points within clusters, but upon examination these weights are not found to be meaningful or appropriate. Furthermore, the weighting factors vary depending on the geometric relationship between the point to be estimated and the data points. Ideally what is required is a cluster weighting which is separable from the distance weighting – and need be computed once only for a given sample point configuration.

There is in fact a method to obtain an objective estimate of the clustering effect for each data point. In a random Poisson point distribution in two dimensions, the number of other data points within any annulus around a given data point is proportional to the distance from that point. In three dimensions it is proportional to the square of the distance, and so on. The observed data point distribution will of course deviate from this ideal. If there is a tight local cluster of say 10 points very close to the given data points, there is an excess over the expected number of data points within this distance range; this excess will be approximately 10 if they are very close. If the cluster is looser, then the excess would be proportionately lower. The ideal and observed point distributions can be shown as a Kolmogorov-Smirnov plot, and the Kolmogorov-Smirnov statistic will indicate the degree of clustering – the maximum deviation of actual from ideal data point distribution. The inverse of this statistic is then a simple, objective cluster weighting factor which directly reflects the size and tightness of the point cluster. This cluster weighting has the immediate advantage that it does not require to be re-computed for every point to be estimated. It can be computed once and attached to the data set, and will remain valid unless and until the data point configuration changes, for example through addition of new observations.

5.4 *Estimation of a mean block value*

A block estimate is strictly the mean of the estimate of all points within the block. There is no need for any special definition as is used in conventional geostatistical estimation. There are of course an infinite number of points in a block, but it is sufficient to estimate a moderate number, evenly spaced within the block, in order to obtain a reasonable block estimate. Working in this way, one has an immediate advantage that if one retains all of the point estimates which are used, it becomes possible to change the block size and compute a new set of block estimates without re-running the point estimation program.

5.5 *Estimation of point or block variance*

One of the claims made for linear-model geostatistics is that the kriging estimation method, uniquely, generates estimate of block variance. One of the demands of geo-statisticians is that any alternative estimation method proposed must 'also' be able to do this.

First, the claim should be examined. Estimates of block variance with any claim of validity can be generated only by linear kriging. If a more complex nonlinear kriging method is used, it is very difficult to obtain such variance estimates. For example, indicator kriging will yield grade estimates but it cannot easily be made to produce estimates of block variance.

Furthermore, the method by which geostatistics provides block variance estimates is entirely uncontrolled by local data values. It is dependent solely on (a) the overall variogram model and (b) the data point positions. Thus a block is given exactly the same variance estimate whether the nearby data used to estimate the block grade are highly variable or are all very similar. This cannot be right. It is clearly a reflection of the inappropriateness of the geostatistical model and its assumptions of stationarity and homoscedasticity.

Second, it is actually possible to obtain block variance estimates from almost any estimation method. If grades are estimated at multiple points within a block, these may be averaged to give a block mean estimate. They may also be used to obtain a variance of the mean – and hence a block variance estimate may be derived. This can be done using simple inverse-power weighted moving averages, or point kriging, or nonparametric estimation methods as proposed in this paper.

6 ANOTHER OPTION: NATURAL NEIGHBOUR METHODS

We may decide that in a particular situation statistics of any flavour has little or nothing to offer us. In such a situation, is there a more rigorous and plausible way for a miners to estimate their asset base? Well, yes, and we can deduce what it is.

Starting from an understanding that ore deposit samples occur where they do as a result of complex ore forming processes obeying the laws of thermodynamics, we cannot expect that these samples will exhibit a Gaussian distribution. They are not random even although they may arise from a chaotic system that gives them a random appearance at first glance. Such samples are drawn from an unknown spatial function that is never expected to be truly homogeneous and, indeed, we actively seek the anomalous regions in that function.

Additionally, these are always one-off samples; each point of space can only be sampled once and so we must deal with non-replicate data. These considerations indicate that statistically-based approaches will be unhelpful in reconstructing the geometry of the deposit.

Further, a suitable method should not present a black-box persona to the user, nor should it contain any subjective steps, because these aspects inhibit verification of results (as we all know, a result is not scientific unless another independent worker may test its veracity).

To provide results that are as realistic as possible, a suitable method would have the barest minimum of assumptions and, since we are aiming at understanding the size and content of a solid ore body, we have already assumed that the ore body, at the cut-off grade, has a continuous surface. That is, we are assuming that the ore deposit samples are drawn from a locally continuous spatial function.

Because we are limiting our assumptions to 'locally continuous', the most suitable estimation method must be locally-based. But that most venerable of locally-based methods, distance weighting, is notorious for misinterpreting irregularly spaced samples, especially if they are clustered. However, such distance-based weighting does give acceptable results for univariate data; this is a clue. It suggests that an appropriate weighting for bivariate data would be area-based and, analogously, a weighting for trivariate data should be volume-based.

Area-based methods do exist. Triangular facets that approximate topography have been used by miners at least since early in the century (Harding, 1923). The planar facets are functionally equivalent to area-based weighted averages of the three data at each triangle's vertices. Indeed, Isted's method extends this estimation approach to trivariate data (Watson and Philip, 1986). The disadvantage of triangle-based weighting is that three data points do not give a convincing estimate when, as expected for scattered bivariate data, an average of six data points immediately surround an interpolation point.

Obviously, a neighbourhood-based weighting is required. But how does one find the full neighbourhood of a given point? Descartes knew (Okabe and others, 1992, p.7). It is just the data which are 'natural neighbours' of the interpolation point.

Two data are natural neighbours if no third datum is closer to some mid-point of the pair than they are. That is, two data are natural neighbours if some circle through them does not enclose a third datum (if it did enclose the third datum, that third datum would be closer to the circle centre, which is a mid-point, than the other two data are).

And two data are natural neighbours if some circle through them and a third datum does not enclose some fourth datum because the third datum is not enclosed either. So by permutation, the first three data are mutual pair-wise natural neighbours. The data which are natural neighbours of the interpolation point are just the set of data that are pair-wise natural neighbours of the interpolation point.

Figure 1 illustrates that five data points are natural neighbours of the interpolation point X. Such natural neighbour subsets can always be found and provide the most direct information regarding the behaviour of the spatial function at the interpolation point according to our assumption of local continuity.

How are the area-based weights determined? Sibson (1981) has made that clear; it is a straightforward geometrical procedure, both elegant and unambiguous. The weights for all data which are not natural neighbours of the interpolation point are zero. If an interpolation point coincides with a datum, the weight for that datum is one and zero for all other data. That is, the weights always lie in the range zero to one and sum to one. The weight for a given datum goes smoothly to zero as the datum drops from the natural neighbour subset.

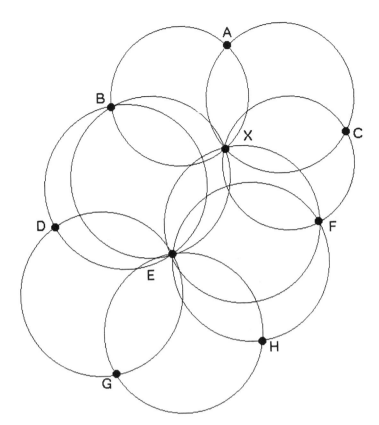

Figure 1. The interpolation point X, has five data A, B, C, E, F which are its natural neighbours, indicated by the natural neighbour circles that it shares with them.

Particularly important, from a miner's point of view, is the fact that these weights, being area-based, are not emulating to a radially symmetric interpolation kernel. This means that irregularly spaced, clustered, or redundant, samples will not bias the resulting surface under the imposition of such symmetry.

The resulting surface is minimal. There is no other surface which includes all the data and has a smaller area – every other surface is more bulgy. So this taut surface is the most conservative interpretation honouring the locally-continuous assumption. There is no set of data for which this method cannot find such a minimal surface so the natural neighbour approach is robust.

But, as pointed out above, this 'rubber sheet' is an extreme or bounding interpretation and, as such, it provides a significant baseline for reserve estimation. Miners should know, as a starting point, the minimal estimate of their deposit under this single assumption of local continuity. The plausibility of the final estimate is more readily judged, both block by block and over the entire prospect, relative to such a baseline.

Of course, such an extreme would usually be too austere to use as a final estimate. Local maxima and minima are modelled as peaks and pits in such a surface and geological knowledge usually will indicate a more rounded, bulbous, treatment. This can be done in a precise and quantitative manner by using the natural neighbour subset to estimate grade derivatives, or gradient planes, for each sample. Then this gradient information can be applied, under user control of range and intensity, to ameliorate the natural neighbour surface yielding a more generous deposit estimate.

Such natural neighbour gradients are not unduly perturbed by an outlier, as a least-squares plane would be, because it is the slope, at that datum, of the taut natural neighbour surface derived from the natural neighbour subset without using that datum, which forms the gradient estimate.

In summary, the natural neighbour method of ore reserve estimation makes only one limited assumption about the data and provides the ultimate in demonstrably conservative estimation in all cases. Clusters of data always are handled correctly and the user has full control of the extent to which the interpretation is augmented by gradients. Although the method is most easily described for bivariate data, it is similarly effective for trivariate data and has been extended to 6 dimensions with success.

Of course, this is not a method for mindless, automatic, use. Sensible application requires a sound geological understanding of prospect boundaries. The number of samples must be sufficient to establish the shape of the deposit to the degree of resolution required. Interpolation must be performed on a grid which is sufficiently dense to portray all the surface complexity inherent in the data (that's the expensive part), but then block averages, grade-tonnage curves, and other statistics can be calculated easily for any subset of grid nodes.

Extrapolation beyond the bounds of the data set, as usual, is highly tenuous, undependable, and problematic, because of course a full set of natural neighbours is not available.

7 THE GEOLOGY

Geostatisticians pay lip service to the importance of geology in estimation of resources. They quite often take care to separate out more or less homogeneous zones for separate geostatistical modelling. However, in reality, the geology should be the principal controlling factor in estimating resources. Real rocks are not random functions, and usually are not well modelled by random functions. Using the straitjacket of geostatistics leads to the situation which is all too familiar to anyone who has estimated gold resources: the same weigthing factors are applied regardless of the actual grade of an actual sample. This can lead in the worst cases to serious over-estimation of poorly sampled parts of a deposit simply because of isolated high-grade samples. This is the true nugget effect. The solution to this problem is intuitively obvious – in such a deposit, high-grade samples should have a much smaller effective range of influence than low-grade samples. This reflects the real geology of the deposit.

Therein lies the key to the whole question of resource estimation. The geology of the deposit should be the real factor controlling the estimation. Often there is insufficient information available, and relatively crude methods such as natural-neighbour, nonparametric methods – or even geostatistics – might be used to give some information. However, where there is some knowledge of the geological setting of the deposit, and particularly where its mode of origin is understood, it should be possible to do modelling in the full sense of the word. A deposit model should ideally take into account the effects of geological processes on the structure and geometry of the deposit and on the localisation of minerals within it.

There is active research today on modelling for predictive exploration (Henley, Ord, and Walshe, in press). This is based on the coupling of geomechanics, fluid flow, and geochemistry in a complex numerical model, to provide a picture of likely mineralisation sites. A similar approach can be adopted for modelling actual ore deposits. Even if a full geological model may be unrealistic for a particular deposit, nevertheless it should be possible to apply some geological intelligence – for example to model the streakiness of an alluvial deposit, or the folding of a Mount Isa or Broken Hill, or the nuggety nature of many gold-bearing veins. Conditional simulation is a technique used in geostatistics, but it misses the point; the conditioning is statistical rather than geological.

A geologically based resource modelling approach will not be a black box. Many of the more mathematical practitioners of geostatistics will not be qualified to use such methods. However, this raises the important question whether such geostatisticians should be considered qualified to do resource estimation at all!

8 CONCLUSIONS

Geostatistics has been applied very widely to estimate the spatial variation of data in many fields, but particularly in mineral resource estimation. Unfortunately, many of its practitioners have lost sight of the fundamental properties and constraints of the geostatistical model, and many of the theoreticians have wilfully violated them in developing new techniques.

Unwarranted assertions of robustness of the geostatistical model risk bringing disrepute to the whole field if users are led to believe that its own assumptions can be violated with impunity. One of the problems is that estimation errors usually are not detected until the deposit is mined, months or years after an estimate has been completed. Errors and discrepancies can often be explained away by a variety of factors including poor prediction of the geology, poor quality of production records, inadequacy of sampling, or many other factors. The geostatistical model itself is held blameless as it provides supposedly the theoretically 'best' estimation method.

Even if the linear geostatistical model is valid in a particular case, it can only provide the 'best linear unbiased estimator' – thus there is potentially an infinitely large family of nonlinear methods which could be better. However, in most cases the model itself is, anyway, inappropriate because of violations of one or more of its basic assumptions. To use derivative methods such as universal kriging or indicator kriging, whose validity is itself questionable or non-existent, merely compounds the problem.

What is required is a more appropriate estimation model, which requires a minimum of assumptions. Since it is generally accepted, even by geostatisticians, that geological data are not even weakly stationary (David, 1977, p.114) it is clear that an appropriate estimation model should not require stationarity. Similar, to handle variables which are not additive, the assumption of additivity (and with it the linear statistical model) must be dropped in favour of a more appropriate model. This paper has outlined three possible approaches: one using the methods of nonparametric statistics, a second providing an objective, repeatable, conservative and non-statistical estimation method using natural neighbours, and the third offering geology and geological processes as the bedrock of any resource estimation.

27

It is not expected that geostatistics will disappear overnight: the geostatistical industry has become too well entrenched. However, a more conscious and cautious use of the methods is required, pending the development of new and more appropriate models to deal with real geological data. And serious efforts are required to develop and validate new standard methods which can deliver what the mining industry requires: reliable and valid resource estimation.

REFERENCES

David, M., 1977. *Geostatistical Ore Reserve Estimation.* Elsevier, 364pp.

Deutsch, C.V., 1994. *Kriging with strings of data.* Math.Geol., 26 (5), p.623-628.

Harding, J.E., 1923. *How to calculate tonnage and grade of an orebody.* Engineering Mining J., 116(11), p.445-448.

Henley, S., 1981. *Nonparametric Geostatistics.* Applied Science Publishers, London, 145pp.

Henley, S., Ord, A., and Walshe, J., in press. *Predictive Exploration: Modelling the Hydrothermal Process.* APCOM, London, April 1998.

Krige, D., 1951. *A statistical approach to some basic mine valuation problems on the Witwatersrand.* J.Chem.Metall.Min.Soc.S.Afr., v.52, p.119-139.

Matheron, G., 1962. *Traite de geostatistique appliquee.* Tome 1 (1962) 334pp., Tome 2 (1963) 172pp., Editions Technip, Paris.

Okabe, A., Boots, B., and Sugihara, K., 1992. *Spatial tessellations: concepts and applications of Voronoi diagrams.* Wiley & Sons, New York, 532p.

Shurtz, R.F., 1997. *Propagation of error and paradox in geostatistics.* R.F.Shurtz, private publication.

Sibson, R., 1981. *A brief description of natural neighbour interpolation.* In Barnett, V., ed., Interpreting multivariate data. John Wiley, p.21-36.

Watson, D.F., 1992. *Contouring - A guide to the analysis and display of spatial data.* Elsevier, 321pp.

Watson, D.F. and Philip, G.M., 1986. *A derivation of the Isted formula for average mineral grade of a triangular prism.* Math. Geol., 18(3), p.329-333.

EQUIPMENT MANAGEMENT – BREAKTHROUGH MAINTENANCE STRATEGY FOR THE 21ST CENTURY

Steven A. Tesdahl
Andersen Consulting, LLB

Paul D. Tomlingson
Paul D. Tomlingson Associates, Inc.

ABSTRACT: Maintenance, in terms of this century, means keeping equipment running or restoring it to operating condition. However, the 21st Century will usher in a broader need for equipment management, a cradle to grave strategy to preserve equipment functions, avoid the consequences of failure and ensure the productive capacity of equipment. Profitable future mining operations will have significantly reduced the 35% of operating costs typically spent on maintenance and the unfavourable impact of downtime that often multiplied these costs by 300%. They will survive those operations who tried to carry outdated 'maintenance' thinking beyond the year 2000. These operations will have applied modern management techniques, technology and information to align the efforts of people with the needs of equipment. Future managers will use equipment management as an integral part of an overall production strategy. They will see it as a direct means of raising productivity, improving performance and minimizing downtime to maximize profitability. This global thinking will characterize all operations planning to survive in the competitive world of mining of the next century.

KEYWORDS: equipment management, culture, reliability centered maintenance, total productive maintenance, technology, information

1 CORRECTING 20TH CENTURY MISTAKES

Not so long ago, the local 'folk' hero in most mining operations was the maintenance foreman. He could bail the operation out of most 'disasters' and manage to get equipment running again. The fact that his omissions contributed to the disaster was never questioned. What mattered was that the equipment was running again. Managers spent little time worrying about how maintenance got done, if at all. Often, there was intense pressure to meet production targets. Against such circumstances, many ignored the elusive 'necessary evil' of maintenance. They chose instead to take their chances that a good ore body, not retiring excess equipment or having a short haul distance would preclude taking actions to utilize maintenance resources more effectively.

Thinking it would bring tighter control of maintenance, many managers designated

29

business units to control both operations and maintenance. While the newly designated business unit leaders liked having maintenance personnel at hand, they found the burden of related activities, like component rebuilding, inconvenient. Business unit leaders were also still focused primarily on meeting production targets. Thus, there was little improvement of the underlying functions that make maintenance work properly. The experience taught managers that simply putting maintenance in a different organizational configuration guaranteed neither better performance nor lower costs. But, the designation of a business unit was still a delegated action. Few managers took direct actions to cause maintenance to become part of the managers overall operating plan. The fact that maintenance cannot perform its work without full support and cooperation has been apparent for some time. See Table I. But, positive actions to acknowledge and correct this matter will continue to distinguish the profitable mines of the future from the rest.

First conducted in 1977 and repeated in 1996, the survey showed little change except in those organizations that created an overall production strategy and placed maintenance within it. These organizations were also characterized by direct management participation.

Thus, the most fundamental aspect of the successful utilization of maintenance resources becomes apparent. Essentially, maintenance alone cannot guarantee consistently reliable equipment that can, in turn, better assure profitability. Experience has established that maintenance makes a greater impact on profitability when all mining

Table I. Improvement of vital maintenance control elements.

Priority	Control Element	Influence Rating	Degree of Maintenance Control (%)	Control Index	Primary Source for Improvement
1	Labor Productivity	10	30	6.0	Maintenance
2	Material Control	10	20	2.0	Other
3	Leadership	9	70	6.3	Maintenance
4	Workload	9	30	2.7	Other
5	Organization	8	50	4.0	Other
6	Interdepartmental Relations	8	20	1.6	Other
7	Cost Data	7	20	1.4	Other
8	Performance Data	7	50	3.5	Other
9	PM Procedures	7	75	5.3	Maintenance
10	Planning	6	60	3.6	Other
11	Scheduling	5	50	2.5	Other
12	Training	4	80	3.2	Other
13	Maintenance Engineering	4	40	1.6	Other
14	Technology	3	90	2.7	Other
15	Labor Practices	2	20	0.4	Other

Respondents rated the degree of direct maintenance influence over 15 vital control elements to establish priorities for improvement. Next, they rated the degree of control that maintenance had over each element. These two ratings were then multiplied to yield a control index. The results indicated maintenance could influence only 3 of 15 control elements (index of over 5.0). The remaining 12 elements could only be improved with support outside of maintenance. (Tomlingson, 1977, pp.5-7)

departments provide coordinated support and that mining managers cause this to happen in a fully accountable manner.

The biggest mistake of the 20th Century was to assume that maintenance could be a stand-alone force somehow capable of guaranteeing reliable equipment, single-handed.

The biggest lesson for the 21st Century will be to ensure that control steps are taken that acknowledge and correct this mistake. Effective maintenance requires the direct participation of every department and the manager in support for, cooperation with and utilization of maintenance resources.

2 EQUIPMENT MANAGEMENT STRATEGY[1]

Strategic Thinking – Any successful strategy is first built on skilful planning followed by effective execution. Successful military strategists, for example, would never enter the battlefield leaving their reserves in a base camp. Yet, by treating maintenance as a stand-alone force, our 20th Century handling of maintenance was much like the failed military strategist attempting battle without reserves. Therefore, to field a winning combination, mining managers must deploy all of their forces. They must assure that each department contributes to reliable equipment in every phase of the life cycle of equipment.

Equipment Life Cycle – An equipment management strategy first acknowledges that mining equipment passes through a life cycle with different departmental support requirements associated with each stage. See Figure 1. These stages include:

- Selection – Determining equipment that best suits intended utilization and performance requirements versus price, ease of operation and maintenance, reputation for quality, manufacturer support etc.
- Purchasing – Carrying out financial transactions to obtain the best equipment
- Commissioning – Placing the equipment in service
- Testing – Ensuring the equipment meets the requirements of the user
- Operating – Operating the equipment during the production process
- Maintaining – Conducting repair and upkeep of the equipment

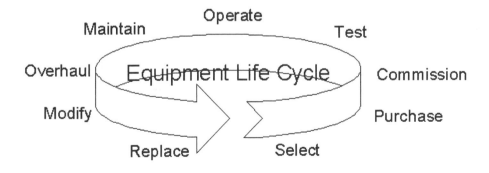

Figure 1. The equipment management strategy covers each stage in the life cycle of equipment.

[1] Doepken, William P., 1977. *Fleet Management Concept.* Amax Henderson Mine. Doepken is now VP, Cyprus Climax Metals, Tempe. AZ.

- Overhauling – Restoring equipment to original design specifications after extended use (may be part of maintaining)
- Modifying – Changing the design configuration of the equipment to yield better performance or to correct or update operating characteristics (a capital expense)
- Replacing – Replacing with similar equipment or improved operating or performance characteristics

Department Responsibilities – (See Figure 2). During the operation stage of the life cycle of equipment, for example, it is obvious that operators operate and maintainers maintain. But, it is less obvious that the personnel department is training new operators and providing higher skill training to maintenance personnel. Similarly, while data processing is recording production statistics and operating costs, accountants and managers are analyzing the information to make fiscal and operational decisions. Concurrently, the warehouse and purchasing are supplying parts and the shops are rebuilding components. Thus, successful equipment management requires the coordi

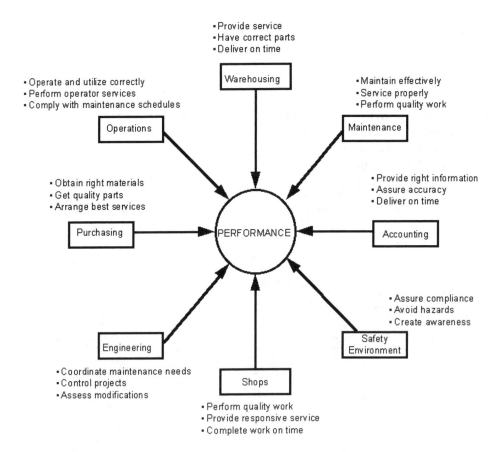

Figure 2. Each department is assigned specific responsibilities that contribute directly to overall equipment management success. Responsibilities are aligned with the appropriate stages in the life cycle of equipment being considered.

nated participation of every department. The strategy pulls their participation together in a cohesive, overall effort. Since various families of equipment will be within different life cycle stages at any one time, the strategy must assign clear department responsibilities for each stage.

Setting Performance Goals – Performance goals are established by managers for each department which, when met collectively, will:
- Assure that equipment has performed effectively
- Require safe, correct operation of equipment
- Guarantee proper maintenance
- Provide for quality material support

These goals put each department into a bottom up performance mode requiring them to establish a high enough standard to meet assigned goals. Performance goals should encourage mutual efforts, typical:
- Purchasing is not getting the right spares unless maintenance agrees that they are.
- Operations are not operating the equipment properly unless maintenance repair history indicates 'zero' problems due to 'operator error'.

Assessing Performance Achievements – At regular intervals, managers meet jointly with departments to review their collective accomplishment of goals and set higher ones. Appropriate information is used to judge the performance of each department. Typical:
- Operations complies with the maintenance schedule 85% of the time
- The warehouse has zero stock outages
- Purchasing delivers materials 24 hours before needed
- Maintenance plans a minimum of 65% of their work

3 GETTING THE FOLKS READY

The Strategy as a Change Agent – The equipment management strategy must also be a framework to change the 'culture' of the total organization toward maintenance. Many may still see it as a 'necessary evil'. Therefore, the strategy also has the task of causing the total work force to see maintenance, instead, as part of an overall operating plan. The strategy must also revise the work force's responsibilities for controlling and carrying out maintenance. There may still be a 'We run it, you fix it!' mentality present in some organizations.

The strategy must create a better appreciation of the strategic importance of effective maintenance. Then, the strategy can more effectively focus people's new thinking and responsibilities on the equipment itself. It will have given people a fresh outlook about maintenance so they can then apply modern technologies and information more effectively to realize the benefits of an equipment management strategy. See Figure 3.

4 A FINAL GOOD BYE TO THE MAINTENANCE CULTURE EXCUSE

Cultural Change – Before any maintenance strategies can be initiated, the attitudes of this century toward the 'culture' of maintenance must be addressed and corrected.

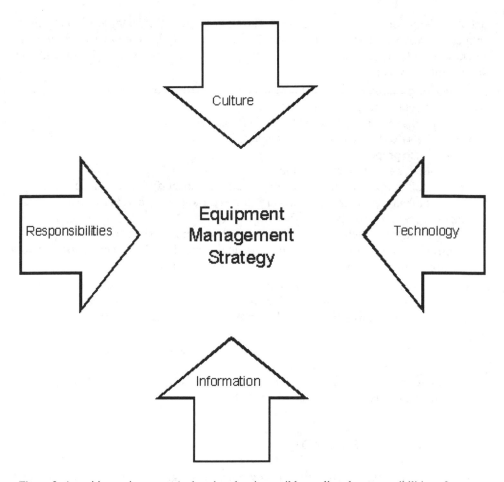

Figure 3. A positive maintenance 'culture' makes it possible to align the responsibilities of personnel toward maintenance and more constructively apply technology and information to implement the equipment maintenance strategy.

Managers who tended to view maintenance as a costly, 'necessary evil' often explained failed improvement efforts by suggesting that the 'culture' of maintenance (knowledge, beliefs and behaviour) precluded their acceptance of potentially beneficial changes. But, we must wonder whether this attitude toward maintenance undermined the outlook of maintenance people and influenced their attitudes toward change, beneficial or otherwise!

Successful implementation of an equipment management strategy must recognize that improving maintenance performance does not begin and end by altering the disposition of 'maintenance' only toward change. Successful implementation must impact the total organization. The need to alter 'culture' must reach out and upwards into operations, staff departments and senior management. All must adjust their thinking to visualize effective maintenance as a means of raising productivity, improving performance, achieving profitability and minimizing downtime. The success of mining operations in improving maintenance must include altering the attitude toward maintenance as a fundamental part of the process of changing maintenance itself.

Change must move from the top down then, the bottom up will respond. Managers must 'create an environment for successful maintenance'. They must ensure that operations and staff departments, like purchasing, understand the need to support and cooperate with maintenance. When maintenance leaders witness 'cultural' improvement toward maintenance, they will display new confidence. In turn, craft personnel will accept and support new improvement strategies. It's that simple!

Efforts to improve maintenance performance have had many successes. Most likely, the successful ones addressed the total organization. Perhaps those who were unsuccessful limited their efforts at 'cultural' change to maintenance itself.

Thus, the 21st Century must include a total organization 'cultural' improvement toward the activity we used to call 'maintenance'. This is a prerequisite to successful application of modern improvement strategies.

5 THE STRATEGY ENCOURAGES TEAMWORK

Reshaping Maintenance Responsibilities – Once the total organization accepts maintenance as an element of the production strategy, the strategy encourages the alignment of responsibilities within a team environment. Typically, teams visualize operators checking, adjusting and cleaning while craftsmen diagnose, inspect, calibrate or replace major components. The organization perceives engineering verifying redesign or modification and purchasing collaborating with maintenance, warehousing and shops to keep the component rebuilding pipeline filled. In addition, management is visualized guiding the overall process with policies and direct participation within a logical strategy.

The harsh, demanding environment of mining can encourage teamwork. Underground coal mining, for example, deliberately makes maintenance personnel part of the operations team. In open pit operations, when a shovel cable must be replaced, no one stands around. In plants, operators move among equipment to check, adjust, clean and make running repairs. Thus, getting people to work together toward the common goal of improving maintenance is already an accepted mode of operation. Therefore, personnel can more readily accept new, well-defined responsibilities for maintenance. These positive working relationships encourage implementation of equipment management strategies. Personnel are now ready to apply their new thinking with programs that address equipment performance directly.

6 FOCUS ON EQUIPMENT PERFORMANCE

What's New – In the past five years, there have been significant people efforts to improve maintenance performance, integrate operations and maintenance, apply new technology and expand maintenance information capabilities. Business units have been created to flatten the organization, implement teams and empower employees to control work in the front line. Operations is now more directly involved in maintenance control than ever before. But, there have also been significant changes in equipment and their maintenance needs. Modern mining equipment must meet more demanding production and quality goals. Therefore, it has been designed for greater reliability. But, it has also become highly complex. Thus, greater technical expertise is required to realize this improved reliability.

The Changing Maintenance Scene – Maintenance organizations are applying new technologies and obtaining better performance information. But, they are also realizing that traditional maintenance arrangements built solely on time-based actions like inspections, servicing, major component replacements and overhauls do not always yield the performance improvements required.

The most promising means of maintaining equipment in mining is not new. Initially developed in 1978 to better assure aircraft reliability, Reliability Centered Maintenance (RCM) has considerable application in mining where maintenance cost and equipment reliability impact profitability so directly. Few other strategies match its potential for attaining maximum equipment reliability and performance while controlling maintenance cost. Reliability centered maintenance brings different but, not new concepts to the maintenance environment. It contradicts the traditional precepts that the reliability of equipment is directly related to operating age. It focuses on preserving the functions of equipment, not on preserving the equipment itself.

Then, by identifying the nature of equipment failures, it specifies actions that reduce the consequences of equipment failure like damaged equipment, possible injury, unnecessary down time resulting in loss of product and, ultimately, reduction in profits.

Team Implementation Effort – The actual task of implementing reliability centered maintenance requires a team effort, especially from operations and maintenance. Operations identifies the functions and the performance standards while maintenance identifies the types of failures. Both collaborate on the consequences of identified failures. Maintenance then defines the most appropriate condition monitoring techniques to preclude functional failure and brings then together in a program. Both then cooperate in carrying out the program, often with specific operator tasks. Thus, the implementing team can attain better understanding of how equipment functions.

In the process, they also gain appreciation of how equipment fails and the root causes. This allows them to better identify specific condition monitoring techniques and apply them to preserve equipment functions and avoid the consequences of failure. The efficient application of advanced condition monitoring techniques is backed up by quality information. This information is used to monitor progress, measure gains, analyze results and record events to identify new applications and benefits of reliability centered maintenance.

The implementation of reliability centered maintenance is a logical progression of eight practical steps that build on shared department responsibilities. See Figure 4.

The implementation steps include:

1. Select the Most Critical Equipment – Let's assume that the most critical equipment in a particular mining operation is the 190 ton haulage truck fleet.

2. Identify the Functions of the Most Critical Equipment – What, exactly do these haulage trucks do in their operating context? The primary function of the 190 ton

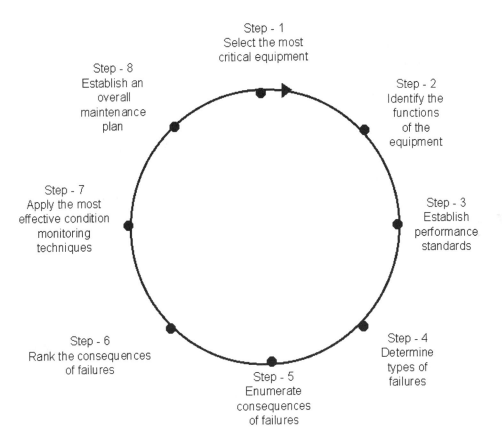

Figure 4. A sequence of eight logical steps is followed to implement reliability centered maintenance.

haulage truck fleet is to move ore or waste from a loading point to a crusher or waste dump.

3. Establish Performance Standards – How well must the haulage truck perform in the conditions under which it operates? Each truck must be able to, for example:
Carry a 190 ton load up and down 12% grades at speeds of up to 25 MPH during all weather conditions for periods of 24 hours, stopping only for refuelling, periodic operator checks and shift change.

4. Determine the Types of Failures – Any equipment condition that does not permit a haulage truck to meet the performance standard would constitute a failure.
A *potential failure* is an identifiable physical condition which indicates that the failure process has started. On haulage trucks, typical potential failures might be:
   • Vibration signalling the onset of transmission failure
   • Cracks indicating the start of fatigue in the truck frame
   • Metal particles in engine oil indicating possible bearing failure

37

A *functional failure* is the inability to meet the specified performance standard. A haulage truck experiencing the following types of failures would not be able to meet its performance standards and would sustain a functional failure:
- Hydraulic pressure is insufficient to raise truck bed
- Low engine compression reduces engine power
- Electrical short inactivates warning devices

Also, we must be aware of *hidden failures* in which the failure is not apparent until the function is attempted by the operator and the truck fails to respond. Typically, these might include:
- The operator pushes the brakes pedals but, the unit keeps going
- The operator activates the lever to raise the bed but, nothing happens

5. Enumerate the Consequences of Failures – What will the result be if a specific failure occurs? Consequences of failure can range from inconvenience to catastrophic. For example, a haulage truck with failed warning lights can be restored to its performance standard with little downtime as the offending fuse is found and replaced. However, a haulage truck without brakes, can pick up speed, collide with another truck, damage both trucks, injure the drivers and block the road for six hours.

   In the larger context, we must consider that maintenance can affect all phases of the mining operation. Typically, without reliable equipment, production targets cannot be met.

   It follows that, without dependable equipment, product quality and customer satisfaction are difficult goals. Then, unreliable equipment can endanger personnel, create environment hazards and even undermine energy efficiency. For all of these reasons, avoidance of the consequences of failure must be a primary maintenance objective.

6. Rank the Consequences of Failures – Because mining equipment has increased in complexity, the number of ways it can fail have multiplied. Therefore, consequences of failure must be classified to guide us in taking preventive and corrective actions. For example:
   - Safety failures endanger personnel as well as equipment
   - Operational failures result in product loss plus the cost of repair
   - Non-operational failures result only in the cost of repair

   In mining, the most important aspects are the avoidance or reduction of the consequences of safety and operational failures. Therefore, the most competent types of preventive and corrective techniques are applied to the equipment most critical to the safety of individuals and the production process.

7. Apply the Most Effective Condition Monitoring Techniques – To detect potential failures early and accurately distinguish them from normal operating conditions, condition monitoring techniques such as vibration analysis are used. They are capable of detecting deteriorating equipment conditions with much greater accuracy and reliability than human beings. These techniques also detect hidden failures that human beings would not be able to detect unless they tried a control mechanism and it did not respond. With the availability of more effective and reliable condition monitoring techniques, equipment condition can be more accurately moni-

tored. This allows a unit to remain in service providing that it continues to meet its performance standard rather than replacing the component at the first sign of potential failure. In turn, this approach yields significantly greater life from components and units.

8. Establish an Overall Maintenance Plan – Based on the consequences of failures, a maintenance program featuring condition monitoring techniques is applied to identify potential failures (starting to fail) accurately and quickly to preclude their deterioration to functional failure (no longer operates) levels. The most effective maintenance program is built on the preceding implementation steps:
   - Critical equipment is identified (190 ton truck fleet)
   - Haulage truck functions are determined
   - Performance standards are established for the truck fleet
   - Types of failures are identified, component by component
   - The consequences of failures are determined for each failure
   - Failure consequences are ranked to give priority to preventive actions
   - The best, most applicable condition monitoring techniques are identified

Then, the condition monitoring techniques selected are fitted into existing, competent maintenance programs to protect the 190 ton truck fleet from functional failures and their consequences.

7 UNDERSTANDING THE EQUIPMENT FAILURE PROCESS

The Failure Process – Tire tread wear is a good example of the failure process. See Figure 5.

Generally, components of any mechanical equipment are subject to wear, corrosion and fatigue. As deterioration increases, the reliability of the equipment decreases. Unless detected and corrected, the deterioration of components increases until the equipment fails. Failures are unsatisfactory conditions that must be considered in the context of the equipment user. A deviation from the performance standard that is un-

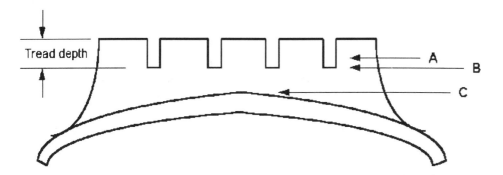

Figure 5. A potential failure is suggested at wear level A. At wear level B, the tire is smooth and further threatens failure. But, if wear continues to level C, the loss is a functional failure. (Nowlan and Heap, 1978, p.31)

39

satisfactory to the equipment user would constitute a failure. But, the difference between unsatisfactory and satisfactory depends on the kind of equipment and the operating environment. To illustrate:

A transmission on a 190 ton haulage truck that leaks oil has failed in the view of the safety department because the oil spillage could cause a fire. However, the same transmission is only a failure to the maintenance engineer when it leaks excessive oil and threatens to fail. But, to the pit production supervisor, the transmission is a failure only when it ceases to drive the truck.

Failures, as in transmission, suggest that the unsatisfactory condition can range from a physical indication that it may stop, to complete stoppage.

So What! – Traditionally, maintenance has observed, detected and corrected failures. They have done this by inspecting and servicing at fixed intervals.

Then, anticipating the age at which components are likely to fail they have replaced them or performed overhauls at predetermined times. This timing often has had no bearing on actual equipment condition. It was simply done at the end of a predetermined period.

But, most failures are not more likely to occur as equipment gets older. Therefore, maintenance programs based on periodic services, component replacements or overhauls are outdated. In fact, overhauls can reintroduce some of the factors that cause the 'infant mortality' failure of equipment.

Condition based maintenance, which carefully monitors actual, current equipment condition, is always less expensive and more effective than fixed interval servicing, component replacements and overhauls throughout the life cycle of equipment (Moubray, 1996, p.21).

Moreover, it is more cost effective to try to improve the way equipment is operated and maintained than to try to redesign it. Redesign should only be attempted if better operation and maintenance won't deliver improvements in equipment performance (Moubray, 1996, p.22).

Mining maintenance, under philosophies such as time-based overhauls, has paid less attention to how components fail and the consequences of failure. There has been an unwarranted assumption that components 'wear out' and become less reliable as operating age increases. Thus, the standard operating procedure in mining maintenance has been to try to restore equipment to an 'as new' condition by periodically replacing components or overhauling the unit. In doing so, maintenance has tended to overlook the failure process itself and the question of what constitutes a failure.

In turn, this omission has led to a maintenance process of avoiding downtime and production loss rather than one based on a wider range of consequences should equipment fail. This helps to explain why we see so much emphasis on meeting production targets and so little (20th Century) attention to maintenance. Thus, reliability centered maintenance reminds us that these consequences impact everything from reliability to profitability and, they demand much more attention than they were getting.

Age Reliability – Reliability is the probability that equipment will survive a definite operating period, under specified operating conditions, without failure. Therefore, the 'life' of a component has little meaning unless a probability of survival is associated with it. Component life (mean time between failure) or failure rates are helpful in budgeting for maintenance tasks established against appropriate intervals. For example:

How many transmissions will be needed for a 12 unit fleet of 190 ton trucks if each transmission lasts about 21000 hours?

But, relative component life explains little about the specific tasks necessary to reduce or avoid the consequences of failure.

Why did transmission in unit 09 only last for 9000 hours causing the loss of the unit to production for 48 hours?

The success of any mining maintenance program can only be judged in terms of how well it prevents the safety or operational consequences of equipment failures. Simply changing components or overhauling units at specific intervals has no impact whatever on avoiding the consequences of failure.

Failure Patterns – Reliability centered maintenance studies have established that there are actually six age-reliability patterns. See Figure 6.

With these failure patterns, mining operations require that maintenance must respond to real needs. Since many failure patterns do not exhibit pronounced wear out periods, maintenance responses must be aimed primarily at detecting potential failures or hidden failures leading to functional failures. Yet, most mining equipment that comes in direct contact with the product (ore, waste, slurry, chemicals etc.), will exhibit definite wear out patterns (mill liners, tires, conveyor belts, drive chains, crusher mantles etc.). Therefore, the maintenance responses must also include removal and replacement of major components within a specified age limit but, only after the exact condition is confirmed with inspection, testing and condition monitoring, never simply at the end of a pre-determined period.

This reality explains why most successful large mining operations have instituted a maintenance engineering effort to more effectively apply and manage the diverse maintenance responses that modern mining equipment requires. Traditional physical inspections, for example, must still be combined with modern predictive techniques like ultrasonic testing. Yet, whatever the responses, they can only be assessed on how well they improve equipment reliability.

Condition Monitoring Techniques and Applications – Condition monitoring is built on the fact that most failures will give some type of warning that they are going to occur (potential failure). It is the physical indication that a functional failure (equipment cannot meet it specified performance standard) is in the process of occurring.

Condition monitoring techniques can obtain precise evidence that a failure is occurring. The condition monitoring techniques used to detect potential failures are called 'on-condition' tasks. That is, equipment is inspected, tested or monitored and then left in service only so long as it continues to meet its specified performance standard. The frequency of these actions to determine whether equipment will be left in service is determined by the P–F interval. It is the interval between the identification of a potential failure and its deterioration into a functional failure.

Basic on-condition tasks are essentially like the human senses. But, while humans can detect problems over a wide range of potential failure conditions, their P–F interval is so short that humans may be watching the functional failure with little ability to provide any lengthy warning period. See Figure 7.

Modern condition monitoring techniques permit potential failures to be detected sooner (the P–F interval is longer). This more sensitive detection of potential failures reveals smaller deviations from normal operating conditions. Thus, more time is available to take the actions to avoid functional failures and consequences.

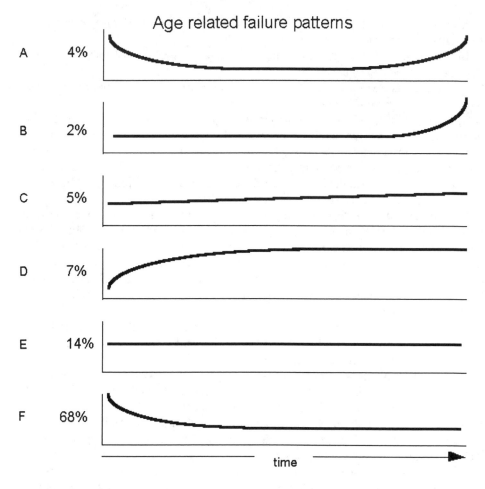

Figure 6. Eleven percent of failures might benefit from a limit on operating age but, 89% do not (Nowlan and Heap, 1978)[2].

Table II illustrates typical effects of equipment operation, the failure symptoms they exhibit and the condition monitoring techniques used to detect them.

Detection and Reporting of Failures – Both operating and maintenance personnel have vital roles in the detection and reporting of failures. Equipment operators observe the dynamic operation of equipment in its normal environment.

They also witness functional failures when equipment fails during use. In addition, they experience the results of a hidden failure when controls fail to respond, for example. Equipment operators are most likely to report the majority of failures because they are on, in or near the equipment during each shift. Therefore, initiatives to involve operators in reporting problems or to perform limited maintenance can signifi-

[2] Although, Nowlan and Heap established these failures patterns in 1978 and they were attributed to aircraft, Moubray seems to have confirmed them in 1994 for a wider range of industrial equipment. His experience was based on a ten year experience in the implementation of RCM in 27 different industries.

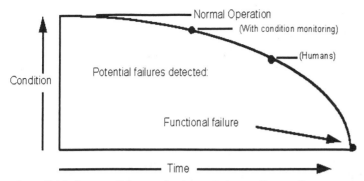

Figure 7. Modern condition monitoring techniques allow earlier, more accurate detection of potential failures and quicker responses. This results in avoidance of more failures and extended equipment life. (Moubray, 1994, p.117)

Table II. Condition Monitoring Effects and Symptoms versus Techniques. (Young, 1995, pp.27-28, and Moubray, 1994, pp.275-6)

Effects	Failure Symptoms	Techniques
Dynamic	Abnormal energy emitted in vibration, pulses and noise	Vibration Monitoring Shock Pulse Ultrasonic Monitoring
Particle	Particles released into operating environment	Oil Sampling Wear Particle Analysis
Chemical	Chemicals released into operating environment	Oil Sampling
Physical	Cracks, fractures, visible and dimension changes	Vibration Monitoring
Temperature	Abnormal increases in the temperature of the equipment (not the product)	Infra-red Thermography
Electrical	Changes in resistance, conductivity, dielectric strength and potential	Infra-red Thermography Surge Comparison Testing

cantly improve overall equipment performance. This reinforces the theme of mutual support of the equipment management strategy.

The operators ability to determine impending failure is further aided by warning devices and instrumentation like on-board computers in haulage trucks that monitor several hundred equipment system performance functions.

Operators and maintenance personnel are an ideal complement in the detection of failures. Operators identify functional failures and hidden failures when controls fail to respond. Maintenance personnel detect potential failures and hidden failures through the effective use of condition monitoring techniques.

8 CONCLUSIONS

The 20th Century has established that 'maintenance' must be an integral part of the production strategy. The 21st Century must build on this. Equipment of the 21st

Century will be more reliable and vastly more complex. Therefore, new strategies and new thinking will be required to realize potential reliability and convert it into profitability. In turn, profitable operations will be those who have used an equipment management strategy to evolve new thinking to successfully apply modern management techniques, technology and information.

REFERENCES

Doepken, William P., 1977. *Fleet Management Concept.* Amax Henderson Mine.

Moubray, J.M., 1994. *Reliability Centered Maintenance.* Butterworth-Heinemann, Oxford.

Moubray, J.M., 1996. *Redefining Maintenance.* Maintenance Technology Magazine, June.

Nowlan, F.S. and Heap, H.F., 1978. *Reliability-Centered-Maintenance.* Office of Assistant Secretary of Defence, Washington.

Resnikoff, H.L., 1978. *Mathematical Aspects of Reliability-Centered-Maintenance.* Dolby Access Press, Los Altos, CA.

Tomlingson, Paul D., 1977. *Industrial Maintenance Management, 4th Edition.* Maintenance Institute, Hofstra University.

Tomlingson, Paul D., 1995. *Mine Maintenance Management, 9th Edition.* Kendall/Hunt Publishing Company.

Young, Keith, 1995. *Integrating Predictive Maintenance.* Maintenance Technology Magazine, June.

I Exploration – Orebody Modelling

GEOSTATISTICAL EVALUATION OF A CHROMITE DEPOSIT

Dimitris Drymonitis

Institute of Geology and Mineral Exploration (IGME), Greece

Keywords: Geostatistics, chromite deposits, ore reserves estimation

A substantial exploration has been made in Greece for the discovery of chromite ore deposits. The continuous evaluation of the explored ore bodies is necessary so that the ore deposit which will be given to exploitation will be economic.

The application of classical evaluation methods, which do not take into account the peculiarities of chromite ore deposits, may lead to wrong conclusions and consequently to wrong decisions for the exploitation.

These are the reasons why geostatistics is used. The application of the right geostatitical method for the evaluation of chromite ore deposits praises the superiority of geostatistics over the classical methods. That is not only because it takes into account the peculiarities of the deposits but it also provides unbiased estimates of the variables and the corresponding errors of estimation.

The data from the deposit under study come from boreholes. The deposit is rather small and low grade and the chromite is of schlieren type. A series of bore-hole sections showed that the body is inclined and subdivided in two zones A and B. The estimation of the reserves in each zone was carried out by using different methods. A Random Stratified Grid (R.S.G.) was fitted to the centres of the intersections of the bore-holes with each zone.

The estimation of the reserves in zone A, which is the most important of the two, was made by using the method known as "unrolling of dipping ore-bodies". This was the most important scientific issue of this evaluation, which had also some other evaluation difficulties, due to the lens shape of the chromite bodies in the zone.

Mailing address: Institute of Geology and Mineral Exploration (IGME), 70 Messogion St., GR-11527 Athens, Greece
E-mail address: ddry@netgate.gr

COMPUTER ORE BODIES MODELLING AND CALCULATION OF DEPOSIT RESERVES AT BURYATGOLD MINING COMPANY JSC BURYATZOLOTO

Munko Dampilov & Victor Vitsinsky
JSC Buryatzoloto, Russia

Vladimir Skagenik
National Polytechnical University of Donetsk, Ukraina

Keywords: Gold mine, computer model, outlining, reserve calculation, complex software

The main functionalities and development principles of the "ARM-Geologist" software complex applied for making the computer reserve model of JSC Buryatzoloto mine are stated.

The software philosophy, milestones of computer graphic model construction, complex application area are described.

The compatibility of software package with the existing packages, as well as its fitness to initial mining and geological information accepted during exploration activity on CIS mines is underlined.

The graphic materials illustrating capabilities of the complex are submitted.

Mailing address: JSC Buryatzoloto, Tsivileva 9, Ulan-Ude, Buryatia, Russia, 670034
E-mail address: root@bgold.buriatia.su

JOINT GEOSTATISTICAL SIMULATION OF CROSS-HOLE TOMOGRAPHIC AND ASSAY DATA FOR OREBODY MODELLING AND MINE PLANNING

Roussos Dimitrakopoulos

WH Bryan Mining Geology Research Centre, The University of Queensland, Australia

Konstantinos Kaklis

Dept. of Mineral Resources Engineering, Technical University of Crete, Greece

Keywords: Conditional simulation, data integration, cross-hole tomography, orebody modelling, mine planning

This paper presents the conditional simulation of orebodies based on the joint use of diamond drill hole assay data and cross-hole tomographic data. The study demonstrates an advanced non-parametric stochastic simulation framework that allows modelling based on the "hard" assay data, while the "soft" tomographic data is used according to their quality/accuracy and compatibility to the assay data. The technique is applied in the modelling of the cross-hole region using Cu assay data and integrating cross-hole RIM tomographic conductivity data to the modelling process. The study shows that modelling based on the direct integration of diverse data is an overly promising and powerful tool for advanced orebody modelling and enhanced mine planning.

Mailing address: WH Bryan Mining Geology Research Centre, The University of Queensland, Brisbane, Qld 4072, Australia
E-mail address: roussos@minmet.uq.edu.au

INTERACTIVE ANALYSIS AND MODELLING OF SEMI-VARIOGRAMS

Gordon S. Thomas

Snowden Associates Pty Ltd, Australia

Keywords: Semi-variogram, ore reserve estimation, kriging, indicator variography, interactive modelling, indicator kriging, mineralisation continuity, software tools

Variography is one of the most time-consuming and user-demanding aspects of geostatistics. Difficulties in visualising and interpreting variography results have been major factors inhibiting the more general application of geostatistical techniques in ore reserve estimation. Such problems are compounded when non-parametric approaches to variography (such as indicator variography) are employed to investigate the spatial behaviour of mineralisation at a range of grade values.

This paper outlines some of the new techniques and tools that have been developed to assist in the generation, analysis and modelling of three-dimensional variograms, with a view to producing high-quality models for use in indicator kriging and conditional simulation. Emphasis is placed on determining whether the low, medium and high grades exhibit similar patterns of mineralisation continuity and whether the directions of that continuity vary with changing grade values.

Mailing address:, Snowden Associates Pty Ltd, Mining Industry Consultants, P O Box 77, West Perth WA 6005, Australia
E-mail address: gthomas@snowdenau.com

GEOSTATISTICS: A CRITICAL REVIEW

Stephen Henley

Resources Computing International Ltd / Snowden Associates (Europe), London, UK

David F. Watson

CSIRO Exploration and Mining, Perth, Western Australia

Keywords: Geostatistics, resource estimation, evaluation, natural-neighbour, non-parametric statistics, geology

Geostatistics has become the accepted standard model for mineral resource estimation. An examination of the underlying assumptions of geostatistics is followed by a review of some theoretical and practical problems in the use of linear and nonlinear geostatistics. Possible alternatives to the geostatistical model are examined, and some less restrictive approaches are presented.

Mailing address: Resources Computing International Ltd, Lutomer House Business Centre, 100 Prestons Road, London E14 9SB, UK
E-mail address: steve@rci.co.uk

ORE GRADE ESTIMATION WITH MODULAR NEURAL NETWORK SYSTEMS. A CASE STUDY

I.K. Kapageridis

AIMS Research Unit, University of Nottingham, UK

B. Denby

University of Nottingham, UK

Keywords: Ore grade estimation, neural networks, radial basis functions, function approximation

This paper introduces a neural network approach to the problem of ore grade estimation. The system under consideration consists of three neural network modules each responsible for a different area of the deposit, depending on the sampling density. Octant and quadrant search is used as a way of presenting input patterns to the modules. Both radial basis function networks and multi-layered perceptrons are used as the building blocks of these modules. An iron ore deposit provides the training and testing data for both the neural network system and kriging, and the results from the two approaches are compared.

Mailing address: AIMS Research Unit, Department of Mineral Resources Engineering, University of Nottingham, University Park, Nottingham NG7 2RD United Kingdom
E-mail address: enxik@unix.ccc.nottingham.ac.uk

TECHNOLOGICAL ORE TYPES IN OREBODY MODELLING NIOBIUM MINERALIZATION IN CATALAO I, GO, BRAZIL

L. M. Sant'Agostino
Geosciences Institute, University of Sao Paulo, Brazil

H. Kahn
Polytechnic School, University of Sao Paulo, Brazil

Keywords: Technological ore types, technological reserves

The ultramafic alkaline complex of Catalao I, located in central part of Brazil, presents a very deep weathering that formed a lateritic mantle with niobium and niobium-phosphate mineralizations associated. The area described in this paper was extensively studied, but it is not being explored because of technological constrictions that reduced the exploitable reserves, compromising the economical feasibility for mining it.

The mineralized material, or the potential ore with high Nb grades, is not homogeneous in its chemical, physical and mineralogical characteristics, with a heterogeneity largely controlled by geological processes.

Ore dressing mineralogy studies led to the definition of several types of potential ores, representing known compartments of the orebody, with distinct characteristics and widely different technological behavior. Some of these types are susceptible to produce the desirable niobium concentrate with good recovery, and others not in the present state-of-the-art for processing this kind of mineralized material, and they were integrated to the tails.

The modeling of the niobium orebody (including the portions with and without phosphate associated) was performed considering these potential ore types, because they present important consequences in the technical-economical feasibility for the deposit exploration.

This paper relates the orebody modeling applying an integrated software - Lynx MMS for workstation - what, even with the tools and facilities of the software for 3D interpolations, was an special task due to the orebody geological setting. The mineralization in this area was partially controlled by supergenic process, with the spatial potential ore types distribution intimately correlated with actual and old topography configuration, geometrically characterized by small and sometimes thin portions.

Coherent result was achieved only with the geological interpretation done by closely spaced planes, meaning interpolation of 10 m thick horizontal sections that were detailed to 4 m thick sections in order to accommodate the distribution of the more superficial potential ore types. This procedure led to a setting that could reproduce the geological configuration deduced from field works and drilling lithological descriptions.

Mailing address: University of Sao Paulo, Geosciences Institute, Department of Economic Geology, R. do Lago, 562 - Cidade Universitaria, 05508-900 Sao Paulo, Brazil
E-mail address: agostino@usp.br

ORE RESERVE ESTIMATION METHOD BY USING GEOGRAPHIC INFORMATION SYSTEMS

Hakan Uygucgil & Can Ayday
Anadolu University, Turkey

Keywords: Geographic Information Systems (GIS), reserve estimation, cross-section method, 3D modelling

In this study, a reserve estimation attempt in a silver ore deposit was carried out by the help of Geographic Information Systems (GIS). Using GIS, 3D model of the studied region, most suitable section planes and grade distributions were obtained and reserve estimation by cross-section method was approached. Besides, tabular cutting depth values were stored, and the relationships were encoded between the drillings, lithological units, and cross-section in a GIS database.

Mailing address: Anadolu University, Satellite & Space Sciences Research Institute, 26470 Eskisehir, Turkey
E-mail address: uygucgil@vm.baum.anadolu.edu.tr

THE CORELOG7 SYSTEM AT ROSEBERY Pb, Zn, Au, Ag, Cu MINE, WESTERN TASMANIA

Robert D. Willis

Pasminco Rosebery Mine, Australia

Keywords: Access, Corelog7, database, diamond drilling, drillcore logging, Pasminco, Rosebery

The Rosebery mine has been operating since 1893. In 1994 the Deep Exploration Project was initiated by Pasminco Ltd to increase a static reserve base. Part of this project involved 100,000m of diamond drilling. A review of the existing core logging and data handling system concluded a relational database product would provide a better system for drillhole data handling. The Corelog7 system is used to log drillholes, manage data, generate reports and export data. Since its commissioning in December 1995, 5600 existing and 800 new drillholes have been incorporated into the system.

Mailing address: Pasminco Rosebery Mine, Hospital Road, Rosebery, Tasmania 7470, Australia
E-mail address: willisr@pasminco.com.au

USING NEURAL NETWORKS TO INTERPRET GEOPHYSICAL LOGS IN THE ZINKGRUVAN MINE. A CASE STUDY

Stefan Wanstedt
GEOSIGMA AB, Sweden

Yi Huang
Lulea University of Technology, Sweden

Lars Malmstrom
Zinkgruvan Mining AB, Sweden

Keywords: Ore delineation, borehole geophysics, drilling, mine planning

The use of borehole geophysics for ore delineation provides the means for a more dense sampling of the ore boundary. Consequently, it is possible to mine ore bodies with irregular geometries without too much dilution. Complex geology and large amounts of information requires elaborate interpretation routines. Here neural networks are used to interpret the logs. This paper shows how data are recorded and interpreted in the Zinkgruvan mine in Sweden.

Mailing address: GEOSIGMA AB, Box 894, S-751 08 Uppsala, Sweden
E-mail address: sw@geosigma.se

II Mine Planning – Mining operations

MINING RISK MITIGATION BY INCORPORATING UNCERTAINTY INTO PLANNING

Joao Felipe Costa, Gonzalo Simarro, Andre Cezar Zingano & Jair Koppe

Federal University of Rio Grande do Sul, Brazil

Keywords: Mine planning, geostatistical simulation, coal mining, gaussian simulation

The profitability of open cut coal mines is extreme sensitive to local reserve variations. Prediction of stripping ratio fluctuation and local reserve are of paramount importance for the success of a mine operation. Estimation techniques such as kriging are used in building block models with less variability than the input data set and referred as smooth models. Kriging or other commonly used methods do not reproduce the spatial variability shown by the data set. Contrary to the smooth models generated by estimation techniques, conditionally simulated coal seams reproduce the actual variability and spatial continuity of the attributes of interest. As a result, simulations provide the tools to address uncertainty in the variability of geological attributes and the resulting effects on various aspects of mine planning and production scheduling. This paper presents the sequential conditional simulation algorithm and illustrates how simulated models can be incorporated into mine planning and scheduling. A case study demonstrates the efficiency of the method in helping reducing risk during mining.

Mailing address: Mining Engineering Department, Federal University of Rio Grande do Sul, Av. Osvaldo Aranha 99/506, Porto Alegre RS-90035-190, Brazil
E-mail address: jfelipe@lapes.ufrgs.br

EOLAVAL - A WINDOWS MINE VENTILATION DESIGN SOFTWARE

Kostas Fytas
Laval University, Quebec, Canada

Pierre Thibault
Quebec Mining Association Inc., Canada

Keywords: Underground mining, ventilation design, simulation of ventilation networks

This paper presents EOLAVAL, an integrated mine ventilation design software. In its new Windows (95, NT) version it is a user friendly fully interactive computer package which can be used to carry out the following aspects of underground mine ventilation design: mine ventilation network analysis and simulation, economic design of an airway, auxiliary ventilation design, calculation of a mine air heating plant, and evaluation of natural ventilation and pressure losses in a mine shaft. This paper concludes with the presentation of a case study illustrating the use of the computer package in mine ventilation design.

Mailing address: Department of Mining & Metallurgy, Laval University, Quebec, G1K 7P4 Canada
E-mail address: Kostas.Fytas@gmn.ulaval.ca

DETERMINATION OF DILUTION OF LIGNITE DURING SELECTIVE MINING WITH BUCKETWHEEL EXCAVATORS USING NEURAL NETWORKS TECHNIQUES

Michael J. Galetakis

Technical University of Crete, Greece

Keywords: Neural networks, dilution, selective mining, bucketwheel excavator, multiple-seam lignite deposits

Dilution of lignite mined from multi-layer deposits during selective mining in open cast mining with bucketwheel excavators is caused by several reasons related to deposit structure, excavation equipment characteristics and mining conditions. Due to the complexity of the problem and the uncertainty of many factors related with it, conventional (traditional) mathematical models are not suitable to describe it and perform a quantitative calculation. Neural networks are especially useful for solving complex problems which are tolerant of some imprecision, which have a lot of training data available, but to which hard and fast rules cannot easily be applied.

A multi-layer feed-forward neural network training by back propagation is developed for the estimation of the dilution of lignite. The trained neural network is used then, to predict the dilution for different mining conditions. Results predicted are reasonably accurate compared with conventional models.

Mailing address: Technical University of Crete, Department of Mineral Resources Engineering, Chanea 73 100, Greece
E-mail address: galetaki@mred.tuc.gr

MINING METHOD SELECTION IN VISUAL DESIGN

Zbigniew J. Hladysz

South Dakota School of Mines and Technology, USA

Keywords: Mining method selection, computerized mine design, 3D visualization, Vulcan.

Traditional process of mining method selection is predominantly based on manual assessment of mineral deposit characteristics. Presently used selection schemes, in which the average values (or the ranges) of the deposit characteristics are compared with those required for each mining method, may be no longer adequate for modern mine design.

The major deficiency of manual assessment is that it does not address the spatial characteristics of the deposit. Today, with the increasing use of integrated computerized design systems, a true, quantitative type of deposit assessment is needed. When mining method selection is performed spatially, its outcome can be visualized in terms of spatial applicability of a given mining method, and the variables interrogated in virtual environment.

The paper presents a new computer visualization technique applied to the process of preliminary selection of candidate mining methods to mine an underground deposit.

Spatial analysis, modeling of deposit characteristics and other prerequisites are accomplished using Vulcan software. Standard computer generated models of the deposit in the form of triangulations and block models are, first, used to generate design grids. The design grids represent the most critical - in the decision making process - factors e.g. deposit dip, thickness and rock strength. The grids provide also a background display that acts as a visual and spatial reference for interactions. The design grids are then compared, through interactive manipulations and graphical Boolean, with those required for each mining method. The best matches are used to select the mining method(s) that is technically feasible. The grids representing the selected mining methods are finally utilized to partition the deposit into zones distinguished by different mining methods and calculate reserves by mining method.

This new technique optimizes the preliminary mining method selection process, indicates - visually and digitally - possible problem areas associated with each mining method, provides cost input for economic analysis and converts - traditionally descriptive and qualitative - mining method selection process into a virtual process with quantitative characteristics.

Mailing address: Department of Mining Engineering, South Dakota School of Mines and Technology, 501 East St. Joseph Street, Rapid City, SD 57701, USA

E-mail address: zhladysz@silver.sdsmt.edu

OPENCAST MINING SYSTEMS MODELLING

Andrija Lazic & Vladimir Pavlovic

Faculty of Mining and Geology, University of Belgrade, Yugoslavia

Keywords: Mining systems, modelling, working environment

The paper deals with basic elements of opencast mining systems modeling which consist of three following stages: working environment modeling, operating technology modeling and production stability testing on open pit mines.

Working environment modeling implies detailed processing of open pit mine real space on the level of bench plane data bases which are formed by interpolation and interpretation of original geological works and all other exploration results. This includes structural properties, engineering-geological properties, physico-mechanical properties, grade properties, chemical, technical and mineralogical testing, as well as all other tests required for complete definition of open pit mine real space. Working environment modeling makes use of interpolation methods and optimization of deposit vertical division into bench.

The second stage of opencast mining systems modeling represents modeling of equipment operating technology on open pit mines, having in view that excavation is the main part of opencast mining systems. Modeling is carried out by complete decomposition (breakdown) of excavation technological processes for all operating and maneuvering operations in line with kinematics and construction properties and operating technological parameters. This model stage also includes a real time image of production systems states and behavior on open pit mines. Use is made of simulation modeling, statistical analysis and Monte Carlo method.

The third stage of opencast mining systems modeling represents the integration of the above two model stages. The real open pit mine limited space is interpreted as a data base containing all essential parameters of the working environment which define mining technological parameters. Simulation technique is used to select equipment operating technological parameters, which yield the required production output, and at the same time, data on excavated material grade are taken.

Production stability is tested on the basis of the span of output and excavated mineral material occurrence in open pit mine real space compared with limit values.

Opencast mining system models may be used as predicative models in the stage of design, as well as normative models for production control during exploitation.

Mailing address: Department for Opencast Mining, Faculty of Mining and Geology, Djusina 7, Belgrade, Yugoslavia
E-mail address: laza@EUnet.yu

THE RELIABLE, EXACT AND FAST PROGRAMS FOR MODELING OF VENTILATION NETWORKS

Alexandr Podolsky

Dnepropetrovsk, Ukraine

Keywords: Mining ventilation, airflow, modeling, account, program

In the paper, problems related to the programs for calculation of airflow in ventilation networks of mines are discussed. The problem of effective modeling of a ventilation network is rather important and requirements to its decision continue to grow. The developers and users of systems of ventilation modeling are compelled constantly to search for a compromise between dimension of the model, accuracy and speed of accounts. The author has developed algorithms and programs ensuring simultaneously high accuracy and high speed of account of ventilation networks. The iterative procedures of the Hardy-Cross method are much modified in view of properties of mining ventilation networks and software and hardware of personal computers. The developed algorithms serve as a basis of programs which have been used for a long time on several Ukrainian and Russian mines. These means allow to ensure reliable, exact, fast and convenient modeling, designing and optimization of mining ventilation networks. A demonstration version of the program is applied to the paper.

Mailing address: 320070 Moskovskaya str. 29/18, Dnepropetrovsk, Ukraine
E-mail address: ap@model.unity.dp.ua

DISPLAYING HIGH QUALITY MINING ANIMATIONS ON A WEB PAGE

Klaus-Christoph Ritter
Otto-von-Guericke Universitat Magdeburg, Germany

John R. Sturgul
University of Idaho, USA

Keywords: Simulation, animation, Internet, WEB-based animations

Existing Simulation and Animation (S&A) software tools are typically platform dependent and do not particularly lend themselves to cooperative work within either the Internet or an Intranet. This paper shows a way to publish simulation results into Internet or Intranet structures.

Most animations on a WEB site are not of high quality. It is now possible to have extremely high quality animations that can be viewed from a WEB page because of software developed by one of the authors (K-C R). This is shown for an underground mine section that was modeled recently using GPSS/H® for the simulation and PROOF Professional® for the animation. The example shows a mine level using trains to haul the ore from the mining section to the ore bins. The trains are restricted in their movements as there is only one way traffic on all the tracks. In addition, only one train can load from an area at a time. The data used in the program was dummy data and the layout changed from the actual mine where the study was carried out. The animation shown can be speeded up or slowed down, zoomed in or out, and has predefined views.

Mailing address: University of Idaho, Moscow, ID 83844-3024, USA
E-mail address: sturgul@uidaho.edu

3-D SIMULATION OF MINING SYSTEMS USING AUTOMOD

Nick Vagenas

Laurentian University Mining Automation Laboratory, Canada

Malcolm Scoble

University of British Columbia, Canada

Keywords: Simulation, 3-D animation, underground mine, mining method, vertical retreat mining, equipment systems

This paper presents a general view of the PC simulation modeling tool AutoMod by AutoSimulations Inc., USA for modeling conventional and automated underground hard rock mining operations. The 3D capabilities combined with built in English like code make AutoMod a suitable simulation modeling tool for mining applications. The resulting graphical representation enhances the understanding of mining operations and the acceptance of simulation as a viable mine design and management tool.

The paper discusses the development of simulation models and the problems related to the use of AutoMod for modeling underground mining operations. Furthermore, the paper will discuss issues related to interfacing orebody modeling software with a mining system simulation tool. The philosophy behind such integration is to compare the performance of automated mining systems on a particular orebody. The successful fusion of the two software systems (an orebody modeling software and a mining system simulator) would provide an overall mining system representation where the results of changes in technical and economic scenarios could be enhanced by means of 3D visualization and animation.

Mailing address: Laurentian University Mining Automation Laboratory (LUMAL), Sudbury, Ontario, P3E 2C6, Canada
E-mail address: nvagenas@nickel.laurentian.ca

APPLICATION OF THE RICHARD'S EQUATION TO SOLUTION MINING

Caleb Wright & Keith Prisbrey
University of Idaho, USA

Thom Seal
Newmont Gold, Inc., USA

Keywords: Solution mining, Richard's equation, gold leaching

In the U.S., solution mining has emerged as the number two dominant method of mining, next to open pit mining. Solution mining's economic importance emphasizes the need for better models. Our objective was to measure and model lixiviant flow in an example of solution mining, gold heap leaching. The procedure was to leach gold in twenty cm diameter, 300 cm high, column leach tests, and then use these results to calibrate Richards equation for solution flow modeling:

$$Cdt(psi) = div(Kgrad(psi))$$

C is the specific moisture capacity, psi is the tensiometer pressure, K is the hydraulic conductivity, and dt is the partial derivative with respect to time. While the gold ore was leached with dilute cyanide solution, tensiometer readings were monitored at different points within the columns. The use of tensiometers in the columns, and later in full-scale heaps, circumvents hysterisis and scale-up problems associated with Richard's equation, and leads to more accurate modeling of solution flow and gold recoveries. A typical modeling result shows the time history of gold extraction at different locations in the heap, and sets the stage for "what-if" computations.

We concluded that better models coupled with on-line sensors could improve solution management, leach cycle planning, and heap design and operation. At present, there is a relative lack of control in heap leaching. By comparison, in modern Carbon-in Leach mills, operators can monitor and see hundreds of sensor outputs, and can rely on advanced control strategies for more efficient operation. We advocate placing numerous sensors in the heap as a move toward better operation and control.

Mailing address: University of Idaho, Moscow, ID 83844-3024, USA
E-mail address: pris@uidaho.edu

III Rock Mechanics – Excavation Engineering

MULTI-FUNCTIONAL RISK ANALYSIS OF ROCK SLOPE STABILITY

R. Halatchev

Mining Department, Ministry of Industry, Bulgaria

R. Dimitrakopoulos

The University of Queensland, Australia

D. Gabeva

Bulgarian Academy of Sciences, Bulgaria

Keywords: Rock slope stability, Sarma's method, probabilistic modelling, safety factor

This paper presents a new approach for a multi-functional risk analysis of rock slope stability. The approach includes the use of a modified deterministic variant of Sarma's method which is based on pseudo-static analysis. The approach accounts for the arbitrary orientation of the seismic forces in the slope, due to different sources such as blasting operations, mechanical loads, and earthquakes. The variant developed here is extended to heterogeneous slopes using the concept of mean-weighted assessment of the geomechanical properties of the rock mass. The probabilistic modelling of the slope state incorporates an analytical determination of the statistical characteristics of the geomechanical parameters and the use of Monte Carlo simulation in assessing the characteristics of the acceleration factor as a criterion for the slope state. Some intrinsic aspects of the practical application of the approach are revealed regarding the building of a geomechanical model of the slope and mathematical description of the geomechanical properties. A case study illustrates the application of the new approach and conclusions are drown regarding its possibilities for covering a wide spectrum of problems to be solved.

Mailing address: WH Bryan Mining Geology Research Centre, The University of Queensland, Brisbane, Qld 4072, Australia
E-mail address: roussos@minmet.uq.edu.au

MONITORING AND NUMERICAL MODELLING OF AN UNDERGROUND MARBLE QUARRY

A. P. Kapenis

Edafomichaniki Ltd., Greece

A.I. Sofianos

National Technical University of Athens, Greece

C. Rogakis

Dionyssomarble Co. S.A., Greece

Keywords: Underground quarry design, back analysis, numerical modelling, rock mass behaviour

The paper reports an early stage work carried out at a new underground marble quarry. It included monitoring of the roof deflections during the excavation of a 6 m wide 3 m high room using a multiple EL Beam sensor system, and stress analysis of the roof-pillar-floor rock mass using numerical modelling and analytical techniques. The observed behaviour of the experimental site was then accurately replicated using computer simulations. The computer models were further calibrated by correlating the computation predictions with the in-situ measurements during the stage of room widening from 6 m to 9 m. The accuracy of model predictions allows for its use as a rock mass replica and for the prediction of maximum room span and optimum exploitation layout.

From the numerical modelling and analytical results it was found that spans up to 30 m wide are feasible for the specific rock mass characteristics. However, further work is needed for the more accurate establishment of the rock mass deformability properties and its structural characteristics which may cause local instabilities.

Mailing address: Edafomichaniki Ltd., 5A Delfon Str., Marousi, GR-15125 Athens, Greece
E-mail address: a.kapen@civil.ntua.gr

MINE PILLAR CHARACTERIZATION USING COMPUTERIZED SEISMIC TOMOGRAPHY

A. P. Kapenis
Edafomichaniki Ltd., Greece

C. E. Tsoutrelis
National Technical University of Athens, Greece

Keywords: Seismic tomography, mine pillars, rock mass characterisation

The results of seismic tomographic imaging undertaken in heavily fractured pillars in an underground bauxite mine using the room and pillar mining method are presented. Following a detailed discontinuity surveying around each pillar using the compass and tape technique, seismic measurements across horizontal and vertical pillar sections were carried out. A pillar stress analysis was also performed at various stages of mining using numerical modelling techniques.

Field results indicated that the seismic velocity range was more affected by the discontinuity systems present in the rock mass than by changes in applied stress field. A high degree of correlation was observed between the seismic velocity recorded inside the pillars and the geometric characteristics of the discontinuity sets present. Also, correlation of the seismic tomographies with both field data from field structural observations of the pillars and from Distinct Element Analysis indicated that the zones of low seismic velocity in the pillars were in good agreement with those zones where blocks were detached due to stress redistribution during the extraction process.

The results of this work demonstrated the usefulness and importance of seismic velocity tomography in providing both a qualitative indication of the internal structure and of the degree of mechanical degradation of rock masses.

Mailing address: Edafomichaniki Ltd., 5A Delfon Str., Marousi, GR-15125 Athens, Greece
E-mail address: a.kapen@civil.ntua.gr

THE INFLUENCE OF THE MINING EXTRACTION ON THE LINING OF A SUNKEN SHAFT

Henryk Kleta & Ryszard Zylinski
Silesian Technical University, Poland

Keywords: Mining extraction, rock mass deformation, shaft lining

A long-term and intensive underground exploitation caused geodeformatic regions where the thickness of the extracted deposit amounts to several tens of meters. In geodeformatic regions it comes to macro-displacements of large rock masses. The influence of macro-displacements can be observed in detail in shafts, which are assigned, to have shafts pillars in future.

In the paper the problem of safe sinking of a shaft which is exposed to deformation caused by mining has been considered. The authors prepared numerical models for finding possible calculated values of deformation of the shaft lining in the future, after its deepening. Based on the models and computer simulations the authors calculated prognosis of rock-mass deformation and deformation of the shaft lining. On this basis, the authors proposed methods of the saving of the shaft lining in the next years.

Mailing address: The Chair for Geomechanics, Underground Construction and Land Surface Protection, Silesian Technical University, Akademicka St. No.2, 44-100 Gliwice, Poland
E-mail address: rzylinski@rg4.gorn.polsl.gliwice.pl

A NEW APPROACH TO INCORPORATE ORE RESERVES AND MAINTAINABILITY OF PIT SLOPE INTO OPEN-PIT MINING PLANS

Katsuaki Koike & Michito Ohmi
Kumamoto University, Japan

Tomoaki Sakanashi
Kumamoto City Office, Japan

Keywords: Mining plan, ore reserves, slope, dangerous block, safety factor

Consideration of both mechanical maintainability and economical efficiency is required in open-pit mining plan for areas containing continuous fractures and fractured zones. In this paper, a simple method to estimate changes of ore reserves and safety factors with progress of mining excavation is described. This proposed method was tested for two limestone mines in Japan. Ore reserves were represented by the weight of calcium oxide contained in limestone and safety factor for mechanically unstable block was calculated by limiting equilibrium analysis. Appearance locations and configurations of dangerous blocks with low safety factors were shown graphically on the digital elevation model of one study site. The ore reserves and safety factors for dangerous blocks at the final mining stage were estimated for different slope angles and locations of planning panels in the other site. The combination of these two estimation results is considered to be useful for optimum determination of a panel.

Mailing address: Kumamoto University, Department of Civil Engineering & Architecture, Faculty of Engineering, Kurokami 2-39-1, Kumamoto 860, Japan
E-mail address: koike@gpo.kumamoto-u.ac.jp

RESULTS OF COHESIVE ROCK MINING WITH DISK TOOLS WITH HIGH-PRESSURE WATER JET ASSISTANCE ON THE LABORATORY TEST STAND

Krzysztof Kotwica

University of Mining and Metallurgy of Cracow, Poland

Keywords: Rock mining, dense rock, disc tool, hydromechanical mining, high pressure water jet assistance, load of the tool, unit energy

Assumptions for the process of hydromechanical mining of rock using a single symmetrical disc tool along a circle route were made in the article. On the basis of the results obtained from the conducted research, empirical as well as analytical and empirical models were developed, describing the process of rock mining using a high pressure water jet and hydromechanical mining of rocks using a symmetrical disc tool along a circle route.

The research was conducted at a special research station for research on single mining tools. Work conditions there were close to real conditions of the mine wall, simulated for a diameter close to 1200mm. The processes of rock mining both using a high pressure water jet and using a symmetrical disc tool with or without the assistance of high pressure water jets were examined separately.

A computer program was created on the basis of the developed models, thereby enabling calculation of the load of the symmetrical disc tool mining along a circular route and the mining effect during mining for pre-set parameters. Diagrams for selected parameters, obtained with the aid of the computer program and illustrating the most favourable parameters for the operation of the system, are presented as follows: symmetrical disc tool - high pressure water jets with the least weight on the tool, the lowest unit mining energy or the greatest quantity of mined output, and comparison of results when mining with mechanical and hydromechanical methods.

Mailing address: University of Mining and Metallurgy Cracow, Department of Mining Machines, 30-059 Krakow, al. Mickiewicza 30, Poland
E-mail address: kotwica@uci.agh.edu.pl

BEGRAPH - A GRAPHICS POST-PROCESSOR FOR DISPLAYING THE BEM PROGRAM "MULSIM/NL" OUTPUT FILES

Thoma Korini

Polytechnic University of Tirana, Albania

George N. Panagiotou

National Technical University of Athens, Greece

Keywords: Boundary Element Method, graphics display post-processor, longwall mining method

BEGRAPH is an interactive graphics post-processor for the MULSIM/NL program, which produces pseudo-three-dimensional and section plots of the stresses, displacements and energy release components calculated by MULSIM/NL. The standard output data files produced by MULSIM/NL are used as input to BEGRAPH which can handle both coarse and fine mesh data files with a maximum capacity of 150 by 150 fine mesh elements. Coloured or black & white plots are also produced depending on printer/plotter availability. BEGRAPH is written to serve as an alternative graphics utility to the MLPLTPCB post-processor, which is included in the MULSIM/NL original package.

Mailing address: Faculty of Geology and Mining, Polytechnic University of Tirana, Tirana, Albania
E-mail address: korini@fgeomin.tirana.al

OPTIMIZATION OF THE BLASTING DESIGN IN THE EXCAVATION OF TUNNELS IN ROCK

Luiz Carlos Rusilo
Instituto de Perquisas Technologicas, Brazil

Eduardo Cesar Sansone
University of Sao Paulo, Brazil

Keywords: Blasting, excavation, tunnel, rock, software, optimization

The design of blasting operations was based for decades on the adjustment of older successful schemes to the geometric conditions of the new excavation and to the local rock mass characteristics, applying a process of cut and try that is very much dependent of the executive technician's personal expertise. In this case it is very difficult to proceed with simulations of the blasting operation in order to analyze cost and rock break optimization, because there is not applied a well defined calculation method that takes in sense the rock and explosive characteristics. The common practice in drilling and blasting operations in tunnels and galleries, is to use drilling patterns with equal diameter blast holes, except for the central cut hole which has bigger diameter in order to act as a free surface to the detonation, in accordance to the particular cut configuration. This makes the operations easier, reducing bit changes to once and reducing the need of drill steels (bits) storing to only two types. Some researches in drilling patterns with multiple diameters to blast floor and contour holes, shows that it is possible to obtain better rock fragmentation and productivity at lower costs. In this paper a computational methodology is presented that allows blasting design using information like excavation geometry, rock mass geomechanical parameters, explosive technical properties, hole size, type of cut, cost parameters, blasting vibrations restrictions and smooth blasting at the excavation contour. This paper also discusses operational aspects of the two methods and analyzes blasting operation performance parameters like: specific and total charge, specific and total drilling meters, ratio (rock blasted volume)/(meters advanced) and ratio (charge)/(drilling meters).

Mailing address: Rock Mechanics Laboratory, Mining Engineering Department, University of Sao Paulo, Sao Paulo, Brazil
E-mail address: esansone@spider.usp.br

TECHNOLOGICAL CHARACTERIZATION OF SOME ORNAMENTAL STONES IN SAO PAULO STATE - BRAZIL

Antonio Stelin Junior, Eduardo Cesar Sansone & Wildor Theodoro Hennies
University of Sao Paulo, Brazil

Keywords: Ornamental stones, Brazil, characterization, physical properties, correlation, world wide web

This paper, written by the team of the Rock Mechanics Laboratory of the University of Sao Paulo, Brazil, presents partial results of a pioneering research in Brazil, about the technological characterization of ornamental stones (rough stones) used in the industry of the civil construction of the state of Sao Paulo, Brazil. In this paper we present results of the determination of the physical properties: dry density, saturated density, porosity and water absorption of 18 types of ornamental stones. It was studied the correlation among these four indexes, and good results were observed, allowing the proposition of a quick procedure for the evaluation of three other properties starting from a simple test on the determination of the dry density. The results of this research, that it is still in process, are available in a world wide web site.

Mailing address: Rock Mechanics Laboratory, Mining Engineering Department, University of Sao Paulo, Sao Paulo, Brazil
E-mail address: esansone@spider.usp.br

VIRTUAL PROGRAMMING APPLICATION AT ROCK MECHANICS LABORATORY: THE UNIAXIAL COMPRESSIVE STRENGTH TEST

G. G. Uyar, T. Bozdag & A. G. Pasamehmetoglu
Middle East Technical University, Turkey

Keywords: Rock mechanics, laboratory tests, virtual programming, uniaxial compressive strength test, simulation

Virtual programming application has grown to be popular in engineering education. Students gain an additional benefit from software that uses visual representation of a calculated result, because it can be used directly and often interactively as an aid to understanding.

This paper describes the virtual program of Uniaxial Compressive Strength Test, and how it can be applied as a useful tool in the Rock Mechanics course. It also gives the flowchart of the program and resultant figures of the test.

Mailing address: Middle East Technical University, Department of Mining Engineering, 06531 Ankara, Turkey
E-mail address: gulsev@ rorqual.cc.metu.edu.tr

STRESS DISTRIBUTION AROUND LONGWALL MINING. MATHEMATICAL EXPRESSIONS

Vladimir Petros

Technical University of Ostrava, Czech Republic

Keywords: Abutment stress, longwall mining

The paper analyses the possibilities of mathematical definition of stress distribution curve in front of longwalls. Distribution analyses were based on measurements carried out on equivalent models. These stress measurements are analysed as to how they can correspond with in-situ stress conditions in rock massif, and what are their principal influencing factors. To formulate the mathematical relationship of stress distribution in front of a longwall, the difference between two hyperbolic functions defined by three constants is proposed. Other similar functions can be added to this fundamental mathematical function in order to formulate stress distribution in more complicated rock massif structures.

Mailing address: VSB - Technical University, 708 33 Ostrava, Czech Republic
E-mail address: vladimir.petros@vsb.cz

IV Mine Equipment

SIMULATION OF A CONTINUOUS SURFACE MINING SYSTEM USING THE MICRO SAINT VISUAL SIMULATION PACKAGE

Zacharias G. Agioutantis & Antonis Stratakis
Technical University of Crete, Greece

Keywords: Simulation, surface mining, Micro Saint

The continuous mining method, which employs bucket wheel excavators (BWEs), conveyors and stackers, is the principal mining method used in all the lignite mines at the Ptolemais-Amydeon basin, located in northern Greece. Lignite combustion provides about 80% of the electrical energy annually consumed in Greece. This paper presents a simulation study of a surface mine which includes production (excavation) subsystems (BWEs) interconnected to stacking (dumping) subsystems. Input parameters include downtime distributions as well as mean time between failure distributions, which were calculated from actual data. A visual simulation package called Micro Saint which allows fast model development as well as animation was used. Simulations were executed under Windows95 on a Pentium computer. Simulation results match production characteristics.

Mailing address: Department of Mineral Resources Engineering, Technical University of Crete, GR-73100 Chania, Greece
E-mail address: zach@mred.tuc.gr

DETERMINATION OF EQUIPMENT DOWNTIME BY AUTOMATED PROCESSING FOR CONTINUOUS SURFACE MINING SYSTEMS

Zacharias G. Agioutantis
Technical University of Crete, Greece

Keywords: Surface mining, data processing, statistical analysis, availability

A number of open pit mines, employing mainly continuous mining methods, are currently in operation at the lignite basin of Ptolemais-Amydeon in northern Greece. This study presents the design characteristics and implementation of a software system developed to facilitate recording and processing of downtime data. This software system runs on a network environment (Novell) on a 24hour basis and is designed to process data entered by the operator. Statistical results are presented per equipment unit including mean values and the standard deviation for various downtime groups as well as operational modes.

Mailing address: Department of Mineral Resources Engineering, Technical University of Crete, GR-73100 Chania, Greece
E-mail address: zach@mred.tuc.gr

EFFICIENT AND PRECISE DATA COLLECTION OF MACHINE PERFORMANCE

Detlef F. Bartsch

O&K Orenstein & Koppel Inc., USA

Keywords: Equipment performance, data collection, objective and accurate analysis

The evaluation of machine performance in a designed application is of equal importance to both the owner/operator of the unit and to the machine's manufacturer. Performance measurement enables the manufacturer to properly adjust the equipment and effectively design future enhancements while the owner/operator may alter the operations/work environment for increased productivity.

Analyzing performance measurement data, identifying improvement areas and implementing both strategic and tactical operational changes increases the competitive advantage of producers in their respective industries.

Performance of equipment in the mining and similar industries may be simply described as achieving maximum production in a minimum amount of time while being reliably performed over a scheduled time of operation.

Mailing address: O&K Orenstein & Koppel Inc., 8055 Troon Circle, Suite A, Austell GA 30168-7849, USA

E-mail address: defraba@atl.mindspring.com

OPTIMISED COMBINATION OF MAINTENANCE TYPES

Johan Gouws

Rand Afrikaans University, Johannesburg, South Africa

Leonie E. Gouws

Melikon Engineering Consultants, Midrand, South Africa

Keywords: Plant and system maintenance, reliability-based maintenance (RBM), reliability-centred maintenance (RCM)

Presently, four main types of plant and system maintenance are commonly used: corrective, interval-based preventive, condition-based preventive, and specialised preventive (or proactive) maintenance. Since each of these methods has specific advantages, but also disadvantages, it is important to search for the optimum combination of the different techniques, for each specific application. The process of finding and implementing such an optimum combination is commonly called reliability-based (or reliability-centred) maintenance. This paper briefly defines the different maintenance types; and then proposes typical combinations of these for the different life-cycle phases of a system. It is highlighted that modern technology can be utilised much more in order to move away from traditional corrective (reactive) maintenance, to more preventive (proactive) maintenance.

Mailing address: Research Group for Dynamic Mechanical Systems, Rand Afrikaans University, P.O. Box 524, Auckland Park, 2006, South Africa
E-mail address: jg@ing1.rau.ac.za

COMPUTER-AIDED ANALYSIS OF DYNAMICS OF BELT CONVEYOR WITH AUTOMATIC FOLLOW-UP DEVICE STRETCHING THE BELT

Roman Jablonski & Piotr Kulinowski

University of Mining and Metallurgy, Cracow, Poland

Keywords: Belt conveyors, tension systems, simulation tests

The paper deals with the dynamics of belt conveyors with the mechanical automatic follow-up device stretching the belt by two stretching cars in the loop unit. The mathematical, discrete model of the belt conveyor is used for the analysis of this system. In the analysis, the belt is substituted by a rheological model describing dynamic properties and a Rayleigh model as a multi-mass discrete representation of the belt continuity. The system of differential equations describing the belt conveyor model is given as a multivariate computer program.

The calculation results gave the following information: tension variations in the belt, acceleration, velocity at chosen points of the belt and displacements of stretching cars. Starting and stopping of the belt conveyor have been considered.

Simulation tests have been experimentally verified on the real object. It was an underground coal-mine belt conveyor (length: 910 m, power 2x150 kW). Authors used a TV-camera to register displacements of the stretching car during starting of the conveyor. The results of simulations and experiments were compared and show a satisfactory similarity.

Mailing address: University of Mining and Metallurgy, Department of Mining Machines and Waste Utilization Equipment, Al. Mickiewicza 30, B-2, pok.6, Cracow, Poland
E-mail address: kulipi@uci.agh.edu.pl

89

OPEN SYSTEMS STANDARDS FOR THE MINING INDUSTRY

P.F.Knights
Catholic University of Chile, Chile

L.K.Daneshmend
Queen's University, Canada

Keywords: Mining, open systems, mine management, data warehouse

Key issues in managing information systems in the minerals industry will likely continue to be those relating to the integration and interoperability of disparate systems. This paper argues that, in order to harness the promised benefits of information technologies, the mining industry should adopt a set of open systems standards to define data formats and protocols for seamless data exchange. This is in direct contrast to the present situation where mining equipment, software and instrument suppliers rigidly adhere to proprietary standards for fear of losing competitive advantage. Benefits to the mining sector resulting from the adoption of such standards would be: the provision of near-time data for executive decision support, and the freedom to choose best technologies. The principal benefits to the leading mining software suppliers will be increased market share as a result of "captive sites" converting to open systems standards. Examples are given of open systems standards developed for related industries such as the Petroleum Open Systems Corporation (POSC) and the Machinery Information Management Opens Systems Alliance (MIMOSA).

Mailing address: Mining Centre, Catholic University of Chile, Vicuña Mackenna 4860, Santiago, Chile
E-mail address: knights@ing.puc.cl

SIMULATION MODELS OF BUCKET WHEEL EXCAVATOR WORK TECHNOLOGY

Vojislav Krstic & Sasa Stepanovic
Belgrade University, Yugoslavia

Keywords: Opencast mining, simulation model, bucket wheel excavator

Simulation models of bucket wheel excavator work technology were developed to define its technological parameters in a block, flank and working trench. They are based on coordination between kinematics-constructive properties of selected equipment and the working environment of the relevant exploitation area, in order to determine the optimum technological parameters that represent a prerequisite in providing the maximum production capacity.

When selecting a bucket wheel excavator, the technical properties of which are presented in form of sequential datafile and are a part of the database on equipment at coal open pits, its kinematics-constructive properties are being taken over. The defined slope angle values, obtained through analysis of slope stability, are transformed and the results are interpreted in function of both a bench height and a legally regulated factor of safety, written in form of the data file.

For simulation models of bucket wheel excavator work technology, a program, namely a program packet Tehno-CAD, in CAD graphic environment was developed. A program for calculation of the technological work parameters of bucket wheel excavator in a block, flank and working trench was developed in the BASIC programming language. A data file of technological work parameters, resulting from calculations, is used by programs developed in the AutoLISP programming language that are applied as a graphic interface when presenting technological parameters of bucket wheel excavator in CAD graphic environment.

Mailing address: Faculty of Mining and Geology, Belgrade University, Yugoslavia
E-mail address: voja@rgf.org.yu

A NETWORKING COMPANIES APPROACH TO PHASED AUTOMATION IMPLEMENTATION IN SOUTH-AFRICAN MINING OPERATIONS

Matthew Loxton
Digital Networking Systems, South Africa

Keywords: Networking, South-Africa, tramming, video, automation, radio frequency, underground networking

The drive towards safer, more productive, and lower cost mining operations sometimes necessitates the use of automation. This can often be implemented in a phased approach using standard networking and IT equipment and software interfacing to industry-standard control and instrumentation (C&I) devices. However, due to lack of focus in this area, the IT industry is often in a poor position to offer solutions for mining companies.

Nowhere is this more true than in the South-African mining industry where years of high mineral prices and relatively cheap migrant labour led to an over-reliance on manual methods over the technology. At the same time, IT and networking companies have been focused on the commercial sector because of the relative ease of solutions and understanding of customer problems. Added to this situation was the disparity in many mines between the C&I and IT departments.

In order to address this, it was necessary to "re-tool" the in-house thinking, and spend more time in understanding the unique difficulties faced by mining companies, and specifically, in understanding the physical and logistical issues encountered in mining operations.

The operational areas identified for further study were:
- Extending the reach of core informatic systems to areas not previously covered, such as underground stores, offices, and workshops.
- Extending voice communications and video feed from working areas such as haulage's, stopes, and draw-points.
- Automation projects for metallurgy and processing plants, power monitoring and control, ventilation monitoring and control, and tramming.

The project chosen for this study is in the haulage's of an underground mine and the purpose is to address the tramming mechanism of ore from an ore-box to a crusher tip. The current method is manpower intensive and has inherent safety and productivity flaws, but is the best "manual" method employable. This situation is both a challenging target as well as having positive effects if a solution could be found.

The preliminary study uncovered that operational areas of this mine are "infrastructure-hostile" and standard copper or fibre based cabling techniques would prove to be problematic. However, by using standard spread-spectrum digital radio-frequency equipment and Multimedia Access Devices, it would be possible to reduce the staffing requirement, improve inherent safety, increase productivity, add accountability for productivity, and pave the way for fuller automation in the future by simply adding functionality.

The possibility exists in the future to carry the automation process to the point at which human interaction is remotely achieved, and only periodic maintenance or breakage's need direct human intervention. This would allow withdrawal of some environmental services such as ventilation and a proportion of hoisting services.

The programme thus addresses most of the prime directives of a mining operation:

- Production. Increased tramming tonnage.
- Cost Reduction. Fewer stoppages, lower staffing.
- Expense Avoidance (none directly identified).
- Safety. Taking staff "out of harms way", creating more direct link to operators.

Mailing address: Digital Networking Systems, P.O.Box 69134, Bryanston 2021, South Africa
E-mail address: matthew@dns.co.za

FORECAST OF BELT REPLACEMENTS USING THE "PROGNOZA" PROGRAM

Leszek P. Jurdziak & Monika Hardygora
Wroclaw University of Technology, Poland

Keywords: Conveyor belts, forecast of replacements, Monte-Carlo method, simulation

The big scale of continuous transportation in open cast lignite mines and high costs of conveyor belts force mines to manage them rationally. The ample statistical material gathered in mines' databases is not utilized sufficiently enough. The PROGNOZA program described in this paper is a proposal for using results obtained by advanced probabilistic methods and data on the operating time of conveyor belts to simulate forecasts of belt replacements for a given period. By means of this program, one can do both detailed simulations (replacements of the particular belt sections) and quick approximate simulations (the number of replacements and the length of replaced belts for individual belt conveyors). The obtained forecasts can be used for the evaluation of different replacement strategies. The program was written in TurboPascal v.7.0, it uses dynamic data structures and is equipped with its own statistical module and a generator of random numbers, including the Weibull distribution suitable for describing conveyor belts' operating time.

Mailing address: Institute of Mining Engineering, Wroclaw University of Technology, ul.Teatralna 2, 50-051 Wroclaw, Poland
E-mail address: mhard @ lns.ig.pwr.wroc.pl

PLANNING AND SCHEDULING OF MINE MAINTENANCE OPERATIONS

Adrian M. Salee

Salee Management Systems, USA

Keywords: Equipment, maintenance, preventive maintenance, predictive maintenance, management, equipment failures, repairs, resources

The systematic planning and scheduling of mine maintenance work permits and supports cooperation between Operations and Maintenance in maintaining mine operating equipment, thereby supporting mine operations. Safety and resource management are enhanced, equipment availability increases and costs reduce. Why then isn't everybody doing planning and scheduling, and how do planning and scheduling get started?

Most non-users of planning and scheduling are content that their current maintenance procedures are adequate; after all, they are operating. But they don't recognize their cost in lost equipment availability and production. Work not planned and scheduled usually takes about three out-of-service hours for one hour on the same job done as planned and scheduled. This applies equally above and below ground, but limitations on the number of pieces of equipment and space available below ground magnify the effect.

Starting and continuing a planning and scheduling system requires realizing its worth, and then forming a commitment to do it. Only then can the needed organizing, procedure development, training and actual operation begin and become successful. But if the commitment is there, the procedures will work and the cost and performance benefits will become realities.

Mailing address: Salee Management Systems, USA
E-mail address: amsalee@aol.com

EQUIPMENT MANAGEMENT. BREAKTHROUGH MAINTENANCE STRATEGY FOR THE 21ST CENTURY

Steven A. Tesdahl
Andersen Consulting LLB, USA

Paul D. Tomlingson
Paul D. Tomlingson Associates, Inc

Keywords: Equipment management, culture, reliability centered maintenance, total productive maintenance, technology, information

Maintenance, in terms of this century, means keeping equipment running or restoring it to operating condition. However, the 21st Century will usher in a broader need for equipment management, a cradle to grave strategy to preserve equipment functions, avoid the consequences of failure and ensure the productive capacity of equipment. Profitable future mining operations will have significantly reduced the 35% of operating costs typically spent on maintenance and the unfavorable impact of downtime that often multiplied these costs by 300%. They will survive those operations who tried to carry outdated 'maintenance' thinking beyond the year 2000. These operations will have applied modern management techniques, technology and information to align the efforts of people with the needs of equipment. Future managers will use equipment management as an integral part of an overall production strategy. They will see it as a direct means of raising productivity, improving performance and minimizing downtime to maximize profitability. This global thinking will characterize all operations planning to survive in the competitive world of mining of the next century.

Mailing address: Paul D. Tomlingson Associates, Inc., Denver, Colorado, USA
E-mail address: pdtmtc@sprynet.com

V Mine Safety – Training

THE MULTIMEDIA PUBLISHING SYSTEM: AN ECONOMICAL DEVELOPMENT PACKAGE FOR INDUSTRY SPECIFIC MULTIMEDIA TRAINING PACKAGES

Emma Derby & John Jury
C.Y. O'Connor College, Western Australia

Keywords: CML, multimedia, training, vet, flexible learning, open learning, computer managed learning, mining, remote learning, bulletin boards, online technology

Mining Industry training in Australia suffers significant barriers. Many operations are sited in remote arid locations, with temperature extremes being a normal part of the work environment. Regional isolation adds significant expense to training either on or off site. This expense coupled with increasing regulation pertaining to safety and appropriate competency levels has led to that many operators are undertaking training within the operation.

This paper introduces the Multimedia Publishing System (MPS®) developed by the C.Y. O'Connor College of TAFE (CYOC) in Western Australia. This system produces Computer Managed Learning (CML) modules in multimedia format. The MPS® is revolutionary in its ability to produce CML modules at resource levels comparable to print based material.

The discussion traverses an overview from CYOC's historic use of bulletin boards (BBS) and DOS-based CML software to the development of the current Windows-based Multimedia Publishing System (MPS®) and the benefits it offers to the global mining industry.

The thrust of the content focuses on the economies of scale, which can be achieved through the use of the MPS® in producing series of modules, configured into a common format, which is determined by the client.

Mailing address: C.Y. O'Connor College of TAFE, PO Box 498, Inkpen Street, Northam WA 6401, Australia
E-mail address: juryjr@northam.training.wa.gov.au

99

WHIP! ™ A TOOL FOR VIEWING VECTOR DRAWINGS IN NETWORKS APPLICATION IN A MINE SAFETY INFORMATION SYSTEM

Jelko Ferlin

Premogovnik Velenje - Velenje Colliery, Slovenia

Keywords: Internet, Intranet, security, safety, simulation

Autodesk® has recently released two excellent tools (utilities) which allow designers to view, send and share design content (in vector form) over the Internet or Intranet. One of them is "Internet tools" inside AutoCAD 14®, which allow users to prepare vector drawings in the DWF™ format. The other is "Whip! ™", a stand-alone product that allows users to view vector drawings in the DWF format within browser. These mean that there is no need for installing (buying) AutoCAD for viewing drawings. Most popular browsers, Microsoft Internet Explorer® and Netscape Navigator® support Whip.

In spite of a well-organised fire protection and the application of flameproof materials, the fire in the colliery can break out. However, the effective security-information system enables us to perceive the fire early enough to put it out by means of well-organised fire protection and the rescue brigade.

The fire protection system in the Velenje colliery has designed on the basis of simulated fire outburst at potentially dangerous underground locations. When the simulation was completed, we divided the underground with fire protection facilities and ventilation system into fire sectors. We have placed meters into the fire sectors and connected them with the security-information system. The instructions for taking prompt actions have been prepared for each fire sector. Beside texts there are maps, presenting locations of meters, the ways for a safe retreat of workers, magazines for fire protection equipment, etc. Upon the registration of smoke or CO on a certain meter, we are now able to define exactly the fire sector with a fire outburst and proper activities can be started.

The only hindrance in our work is the extensive documentation that has to be examined in order to find the instructions for the jeopardised area. As we prepared the maps with AutoCAD, we as well used AutoCAD for checking and presentation of maps. However, it had taken some time before the drawing was displayed. That is why we have searched for a suitable (fast) drawing display tools.

It seems that with Whip we finally got just right tools. With a little programming experience (Jscript, Javascript or VisualBasic) we can make suitable application that is fast and simple to use. Moreover, we can include all necessary controls and functions that simplify procedures during a rescue operation, into drawing of the jeopardised area. Instead of looking and searching for the right drawings and the corresponding papers, all steps, measures and procedures are "documented" on the screen (within drawing) and displayed just on time.

Mailing address: Premogovnik Velenje, Department of Research and Development, Partizanska 783320, Velenje, Slovenia
E-mail address: jelko.ferlin@rlv.si

NEW APPLICATIONS OF COMPUTER GRAPHICS AND VIRTUAL REALITY IN THE MINERALS INDUSTRIES

Damian Schofield, Bryan Denby & M. Williams
AIMS Research Unit, University of Nottingham, UK

Andrew Squelch
CSIR, Division of Mining Technology, South Africa

Keywords: Computer graphics, virtual reality, risk assessment, accident investigation, process planning, ergonomics, hazard awareness, driving simulators, safety awareness

The larger mining companies of the world are increasingly using computers to visualise complex minerals operations from a variety of different viewpoints. Advances in computing power, and a reduction in the price of computing hardware has allowed minerals companies to use more advanced Computer Graphics (CG) technologies and Virtual Reality (VR) solutions.

The AIMS Research Unit at the University of Nottingham has been involved in the development of this technology for a number of years, developing a range of CG and VR tools. These tools have been widely applied within the minerals industry. This paper will introduce the latest CG and VR work from the AIMS Unit.

The new applications discussed in this paper include :
- A surface mine truck simulator which is controlled by a steering wheel and pedals.
- Virtual systems for teaching inspection procedures.
- New methods of risk assessment using virtual reality techniques
- A new system for analysing the sight lines around Free Steered Vehicles underground.
- A VR simulator for underground hazard awareness training in the South African gold mines.

Practical examples of the application of these systems will be given and this Internet paper will contain images and animations from a number of these systems.

Mailing address: AIMS Research Unit, School of Chemical, Environmental and Mining Engineering, University of Nottingham, University Park, NG7 2RD, UK
E-mail address: enzds@unix.ccc.nottingham.ac.uk

VI Reclamation – Environmental Issues

ENVIRONMENTAL MANAGEMENT OF AN AGGREGATE QUARRY.
A COMPLEX CASE STUDY

D. Kaliampakos, D. Damigos & A. Benardos
National Technical University of Athens, Greece

Keywords: environmental management, environmental impact assessment (EIA) and monitoring, quarry design, visualization

The conflict between people and mining, may sometimes take extreme, even violent, forms, as in the case illustrated in this paper. In such cases, partial remedial actions have been proved to be inadequate. In contrast, investigation of the situation in depth, based on an integrated environmental monitoring system, becomes imperative. Additionally, in order to eliminate effectively the environmental impacts, as well as to change radically the hostile public opinion and establish friendly relations with the local community, a total re-engineering of the exploitation method is often required.

Mailing address: Department of Mining & Metallurgy, Section of Mining Engineering, National Technical University of Athens, Greece
E-mail address: mmesdk@central.ntua.gr

FUZZY-NEURAL SYSTEMS FOR ADAPTIVE REASONING ON ENVIRONMENTAL RISK ANALYSIS

Marcello M. Veiga & John A. Meech
University of British Columbia, Canada

Keywords: Environmental risk assessment, fuzzy systems, mercury pollution, monitoring programs, adaptive reasoning, expert systems, heuristics

Fuzzy Logic can be used within an Expert System to establish a flexible rule-based mapping of multiple variables into the concept of environmental risk. Rules can be added, deleted or modified as required without worrying about interactive effects. The strength of each rule in the system can be adjusted using neural weights which are learned from existing data about the subject in question. These weights can be adjusted as desired to accommodate changing circumstances or volatile domains. The application of this approach to model bioaccumulation risk from mercury pollution will be presented. The ability of this system to handle varying geographical sites, varying political circumstances and varying points in time will be demonstrated.

Mailing address: Department of Mining and Mineral Process Engineering, University of British Columbia, Vancouver, B.C., V6T 1Z4, Canada
E-mail address: jam@mining.ubc.ca

AUTHOR INDEX

KEYWORD INDEX